CONTENTS

GUIDED EXAMPLES IN MATHEMATICAL TOPICS—General Plan

Series I For the 11 to 14 age range. Each book offers 4 to 6 weeks' work in each of two successive years	Series II For the 14 to 16 age range, linking with CSE and GCE 'O' level syllabuses. Each book offers 4 to 6 weeks' work in two successive years
Year 1 and Year 2	Year 3 and Year 4
1: Sets 2: Graphs and Co-ordinates I 3: Statistics and Probability I 4: Motion Geometry and Vectors 5: Number	6: Relations, Functions and Groups 7: Graphs and Co-ordinates II 8: Statistics and Probability II 9: Matrices 10: Computation

GUIDED EXAMPLES ... OPICS

7: GRAPHS AND CO-ORDINATES II

by J. W. Bradley, M.A.

Gener
ALAI
Head o
Bishop

FREI
LOND

Diagrams by F. R. Fraser

ISBN 0 7232 1123 X

Printed in Great Britain by
William Clowes & Sons Ltd, London and Beccles

PUBLISHER'S FOREWORD

This series is intended to be much more than a collection of graded examples (though it is that too). The introduction of new mathematical concepts and the re-presentation of 'old' topics in ways which reflect modern ideas is a task facing most mathematics teachers today and to this end several major courses are available. Often however set 'courses' can impose a more rigid régime and involve more reading for the pupil than, for many purposes, is either satisfactory or desirable; and for smaller schools courses may be too demanding on mathematics staffs to be a practicable proposition at all.

Mr Charlwood's aim in devising this new series with his team of experienced authors has been to overcome these particular difficulties. 'Reading' is cut to a minimum and reliance is placed rather on the method of Demonstration (by providing detailed worked examples with brief linking text) and Guided Action (by supplying plenty of concept-revealing practice problems). The series should be thought of, therefore, as one which introduces new mathematical ideas in a way which is at once flexible, labour saving, and well-controlled. Special features are as follows:

- The worked examples and brief explanatory texts linking the groups of practice problems are designed to expose points of difficulty, to ensure continuity, and to connect the topics being studied to basic mathematical concepts.

- The books can be used entirely separately or as a series. Teachers may therefore choose whichever topics best meet their particular needs.

- Each book offers 4 to 6 weeks' work in each of two successive years. Series I covers the 11 to 14 age range and Series II the 14 to 16 range. Series II links with CSE and GCE 'O' level syllabuses.

- Each book contains notes for teachers so that they may more easily see how the work is planned and how it links with other topics and other levels.

The presentation makes the series suitable for use as an introductory course in its own right, or as a supplement to traditional courses by offering suitably graded examples in new topics, or as a source of additional examples for one of the existing 'modern' courses.

The general plan of the series is set out in tabular form on the back of the contents page.

TEACHER'S NOTES

This book has been designed for use with pupils in their third, fourth and fifth years at secondary school. (Age range approximately 13–16.) The book is to follow *Graphs and Co-ordinates* in Series I, continuing to give a modern approach to graphical ideas, linking the work with sets, inequalities and practical work. Emphasis is placed on the use of metric units. The book covers all the graph work required for CSE and for GCE 'O' level. Chapters One and Two are suitable for half a term's work in year three. Chapters Three to Six can be used for half a term's work in year four for brighter pupils, but would perhaps be best split into two sessions for the less able.

A knowledge of the following is assumed:

The contents of *Graphs and Co-ordinates* in Series I, except the last chapter.
Expansion of brackets.
Simple change of the subject of a formula.
Completing the square (for sketching parabolas).
Solving simple quadratic equations by factors.
Radians for Exercise 19. This may be omitted altogether.

Pupils should be encouraged to keep all their work, as at times reference is made to earlier questions. Revision Exercises are included at strategic points and are suitable for use as homework or tests, but please note that some of the questions are rather long. Throughout the book the size of graph paper envisaged is 7 inches by 9 inches; a smaller size may make the finding of suitable scales difficult. Centimetre graph paper with large squares of size 2 cm divided into small squares of size 2 mm is equally suitable. Pupils should attempt all questions in the book (including the Revision Exercises); special help may be required for the following:

1. Exercise 4 on determination of a linear law
2. Exercise 14 on linear programming
3. Sine, log and exponential curves at the end of Chapter Four

These may be omitted altogether with less able pupils.

Chapter One

DETERMINATION OF A LINEAR LAW

It was illustrated in *Graphs and Co-ordinates* in Series I that all straight lines have an equation of the form $y = mx + c$ where m and c are two constants; m is the gradient or slope of the line and c is the intercept on the y-axis.

Exercise 1

1 On graph paper draw the following lines for $-3 \leqslant x \leqslant 3$. Remember that it is only necessary to work out the co-ordinates of three points.

(a) $y = 2x$
(b) $\{(x, y): y = x + 1\}$
(c) $y = -x + 4$
(d) $\{(x, y): y = 5 - 2x\}$

2 Draw $y = 2x - 1$ for $-3 \leqslant x \leqslant 3$. Where does this line cut the y-axis? Which of the following lines is parallel to $\{(x, y): y = 2x - 1\}$?

(a) $y = 2x$
(b) $y = -2x$
(c) $y = x + 2$
(d) $y = \frac{1}{2}x$
(e) $\{(x, y): y = 1 + 2x\}$
(f) $\{(x, y): y = 2x + 4\}$

3 Without using graph paper, draw sketches of the following lines. Think carefully about the significance of the negative gradients.

(a) Gradient 1, passing through $(1, 1)$
(b) Gradient $\frac{1}{2}$, passing through $(-1, 2)$
(c) Gradient -2, passing through $(1, -1)$
(d) Gradient -1, passing through $(2, 4)$
(e) Gradient 3, passing through $(-2, -3)$
(f) $y = -3x - 2$
(g) $y = \frac{1}{4}x + 2$
(h) $y = -2x + 4$
(i) $\{(x, y): y = -x + 1\}$ (j) $y = 4$

4 If the equation is linear but not arranged in the form $y=mx+c$, an algebraic rearrangement is needed before the line can be drawn.

e.g. $2y-x=4$ is a linear equation (no terms in x^2, x^3, $1/x$, etc.)
$2y=x+4$ (adding x to both sides)
$y=\frac{1}{2}x+2$ (dividing both sides by 2)

Sketch the following lines, first performing any necessary rearrangement to obtain the form $y=mx+c$.

(a) $3y=3x+4$ (b) $y-x=6$ (c) $2y+4x=1$
(d) $y+2x-5=0$ (e) $y=4x-1$ (f) $x=6-y$

Finding the gradient of a given line

Fig. 1

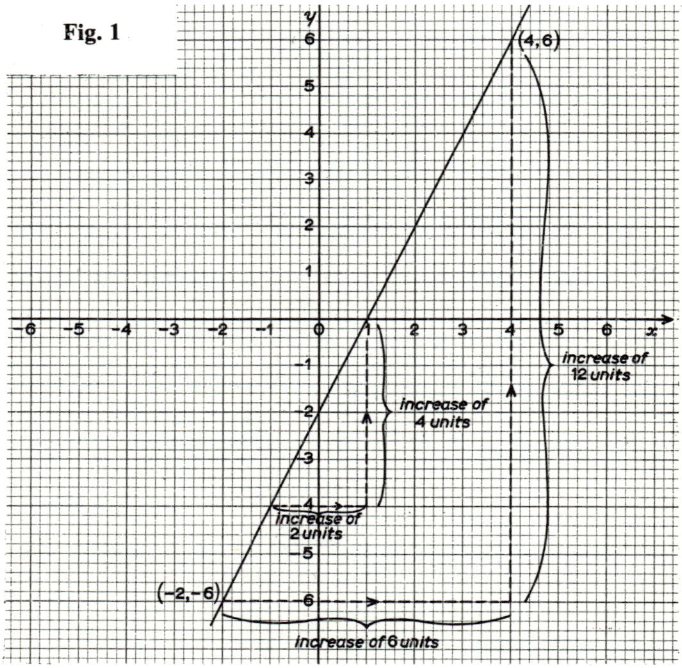

Consider Fig. 1. This illustrates the following question.

Using 2 units per inch on both axes, draw the line joining $(-2, -6)$ to $(4, 6)$. Calculate the gradient of this line.

Now

$$\text{gradient}=\frac{\text{increase in } y}{\text{corresponding increase in } x}$$

so once the required line has been drawn we choose any point from which to start measuring the increases in x and y. You will see this has been done twice in Fig. 1, to illustrate that it does not matter which point is chosen. Imagine first we choose $(-2, -6)$ and decide to increase to $(4, 6)$; the increase in y is 12 units and the increase in x is 6 units.

Hence

$$\text{gradient} = \tfrac{12}{6} = 2$$

If we choose to start with $(-1, -4)$ and increase to $(1, 0)$, the increase in y is 4 units and the increase in x is 2 units, i.e.

$$\text{gradient} = \tfrac{4}{2} = 2$$

Exercise 2

Using 2 units per inch on both axes, draw the lines joining the following pairs of points. In each case calculate the gradient.

1 $(-4, -4)$, $(3, 3)$ **2** $(-3, -6)$, $(1, 2)$

3 $(-6, -1)$, $(3, 0)$ **4** $(-3, 6)$, $(2, -4)$

5 $(-2, 4)$, $(6, 0)$ **6** $(-5, 2)$, $(6, 3)$

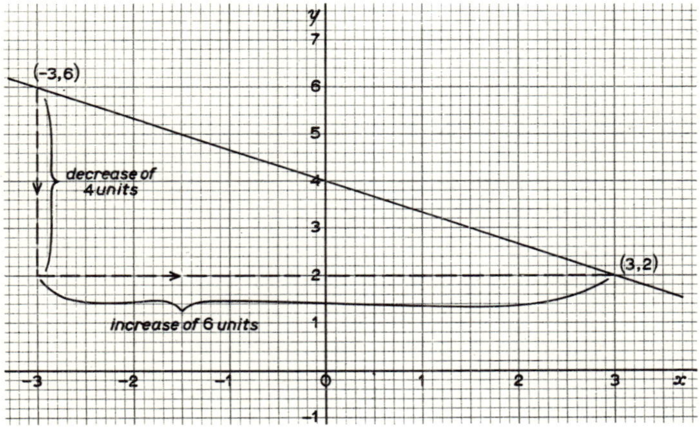

Fig. 2

7 Using 1 unit per inch on the x-axis and 2 units per inch on the y-axis, draw the line joining $(-3, 6)$ and $(3, 2)$. Calculate the gradient.

This question is illustrated in Fig. 2. Note that the amount of increase or decrease in x and y must be measured against the scales on the axes. Hence the illustrated change in y is 4 units although only 2 inches are involved.

Thus $$\text{gradient} = \frac{-4}{6} = \frac{-2}{3}$$

8 Using 1 unit per inch on the x axis and 2 units per inch on the y-axis, draw the lines joining the following pairs of points. In each case calculate the gradient.

(a) $(3, 6)$ $(0, -3)$
(b) $(-2, 6)$ $(1, -6)$
(c) $(-2, -2)$ $(3, 3)$

9 Using 5 units per inch on the x-axis and 2 units per inch on the y-axis, draw the lines joining the following pairs of points. In each case calculate the gradient.

(a) $(5, 2)$ $(20, 4)$
(b) $(-10, 6)$ $(5, -3)$
(c) $(-10, -4)$ $(10, 8)$

Finding the equation of a given line

Look again at Fig. 1 and the question involved. Now we wish to know the equation of the line which joins $(-2, -6)$ to $(4, 6)$. The equation will be of the form $y = mx + c$ where m is the gradient of the line and c is the intercept on the y-axis. The gradient of this line is 2. Thus $m = 2$.

The line crosses the y-axis at $y = -2$. Thus $c = -2$.

Therefore the line is $$y = 2x - 2$$

Similarly for Fig. 2. The gradient of the line is $\frac{-2}{3}$ and the intercept on the y-axis is 4.

Therefore the line is

$$y = \frac{-2}{3}x + 4$$

or $$3y = -2x + 12$$

Exercise 3

1 Look again at all the graphs you drew for Exercise 2 and in each case find the equation of the line.

2 Choosing suitable scales, join the following pairs of points and thus determine the equation of each line:

(a) $(-1, 2)$ $(5, -4)$
(b) $(-20, 1)$ $(40, 4)$
(c) $(-12, 2)$ $(6, -1)$
(d) $(-4, -3)$ $(2, 0)$
(e) $(-6, 2)$ $(3, 2)$

Determination of a linear law from experimental results

When the results of an experiment are known it is usual to plot a graph (or several graphs) of the results in order to see if any generalisations are feasible. The lines of best fit are drawn, and should these turn out to be straight lines the equations may be determined and then conclusions may more easily be reached.

Exercise 4

1 The following are measurements of the speed of a sports car accelerating away from rest.

Time in sec	0	2	4	6	8	10	12	14	16
Speed in km/h	0	13	28	42	54	66	83	96	112

Plot these points and draw the line of best fit. You will find that it is a straight line, and thus there is a linear relation between the speed (s) of the car and the time (t).

i.e. $s=mt+c$ where m and c are constants.

From your graph find the values of m and c. If the sports car continues to accelerate at the same rate, calculate how long it takes to reach 150 km/h.

2 A restaurant manager finds that the cost (C) in £ of running his restaurant depends on the number of meals (M) which are served. The relationship is shown by the following table.

M, no. of meals	50	60	70	80	90	100	110	120	130	140	150
C, cost in £	14	16	18	20	22	24	26	28	30	32	34

Draw a graph to show there is an approximate law of the form $C=mM+c$, where m and c are constants.

If the price of each meal is 25p, how many meals must be served so that no loss is incurred?

3 A ball is thrown vertically downwards with an initial velocity of 3 metres per second. Its velocity v is measured at 1 second intervals and shown in the following table.

Time in sec, t	1	2	3	4	5
Vel. in m/s, v	13	23	32	42	52

Verify that there is a linear law of the form $v=mt+c$ (where m and c are constants). Read from your graph the time taken to reach the following velocities:

15 m/s, 27 m/s, 40 m/s, 60 m/s

Calculate the velocity after 10 seconds.

4 The temperature of the water in an electric kettle was taken at 30 second intervals and the results are shown below. The water boiled after 170 seconds.

Time in sec, t	0	30	60	90	120	150	170
Temp. in °C, θ	15	30	45	60	75	90	100

Plot a graph of these results and thus obtain the equation connecting θ and t.

Chapter Two

PARABOLAS

You may remember that to draw a parabola it is necessary to work out the values that y takes at each integral value of x (and possibly at some non-integral values also). Thus to draw $y=2x^2-4x+1$ for $-3 \leqslant x \leqslant 4$ one first makes a table of values.

x	-3	-2	-1	0	1	2	3	4
$2x^2$	18	8	2	0	2	8	18	32
$-4x$	12	8	4	0	-4	-8	-12	-16
y	30	17	7	1	-1	1	7	17

Thus the points $(-3, 31)$ $(-2, 17)$ $(-1, 7)$, etc., lie on the curve. After choosing a suitable scale these points are plotted and joined by a smooth curve.

Exercise 5

1 Draw the curve $y=2x^2-4x+1$ for $-3 \leqslant x \leqslant 4$. Does this curve have a lowest point? Does it have a highest point? What is the equation of its axis of symmetry?

2 Draw the curve $y=x^2+2x-3$ for $-4 \leqslant x \leqslant 2$. Does this curve have a lowest or a highest point? On the same page draw the line $y=1$. Give the co-ordinates of the points where $y=1$ crosses the curve. Does (1, 0) lie on the curve?

3 Draw the parabola $y=-x^2+2x+1$ for $-3 \leqslant x \leqslant 4$. Does this curve have a lowest or a highest point? If so, give co-ordinates. On the same page draw the line $y=-2$. Give the co-ordinates of the points where $y=-2$ and $y=0$ cross the curve. Give the equation of the axis of symmetry.

4 Draw, on the same sheet of paper, the graphs of $y=x^2+2x-1$ and $y=2x^2-3x+3$ in the range $-3 \leqslant x \leqslant 4$. Write down the co-ordinates of the lowest point of each curve and the points where the curves cross each other.

5 Draw the parabola $y=-2x^2+8x+3$ for $-2 \leqslant x \leqslant 4$. Give the values of x where this crosses the lines $y=0$, $y=5$, $y=10$.

6 Draw the parabola $y=(x+4)(x-2)$ for $-6 \leqslant x \leqslant 4$. Where does this cross the x-axis? What are the co-ordinates of its minimum or lowest point?

7 Draw the parabola $y=2(x-3)(x-1)$ for $-1 \leqslant x \leqslant 5$. Where does this cross the lines $y=0$, $y=-2$, $y=x$? What is the minimum of y?

8 Draw the parabola $y=(5-x)(1+x)$ for $-2 \leqslant x \leqslant 6$. What is the maximum value of y? Where does this curve cross the lines $y=0$, $y=4$, $y=9$ and $y=x+1$?

9 Draw the parabola $y=x^2+x+1$ for $-3 \leqslant x \leqslant 3$. What is the minimum value of y? Give the equation of the axis of symmetry. Does (0, 1) lie on the curve?

10 Draw the parabola $y=(1-2x)(1+x)$ for $-3 \leqslant x \leqslant 3$. Give the equations of the axis of symmetry. State which of the following points lie on the parabola: (0, 1) $(-2, -5)$ (1, 1) $(\frac{1}{2}, 0)$.

11 Draw the parabolas $y=x^2-1$ and $y=1-x^2$. Give the axes of symmetry of these parabolas. Where do they cross the line $y=x$?

12 Draw the parabolas $y=x^2$ and $y=3x^2-5x-3$. Where do these parabolas cross? Give the co-ordinates of their maximum or minimum points.

The position of the maximum or minimum point

You will have noticed that in Exercise 5, question 9, the minimum value of y was not as easy to ascertain as it had been in the previous examples. This was because it did not lie on integral values of x and y. If you look carefully at the tables of values which you worked out for the questions in Exercise 5, you will see that each time the values of y were symmetrical.

e.g. The values of y for question 8 were:

$$-7 \quad 0 \quad 5 \quad 8 \quad 9 \quad 8 \quad 5 \quad 0 \quad -7$$

i.e. Symmetrical about the central no. 9.

In fact $y=9$ was the maximum value for y. In question 9, the values of y were:

$$7 \quad 3 \quad 1 \quad 1 \quad 3 \quad 7 \quad 13$$

Again there is symmetry, but this time about the space between the ones. This implies that the maximum or minimum point of the parabola will lie between the values of x which gave $y=1$, i.e. midway between $x=-1$ and $x=0$.

To find the position of the minimum point it is necessary to work out one more value of y.

Thus $x=-\frac{1}{2}$ gives $y=(-\frac{1}{2})^2+(-\frac{1}{2})+1$

$$=\tfrac{1}{4}-\tfrac{1}{2}+1$$

$$=\tfrac{3}{4}$$

Consider the curve $y=-x^2+3x-2$ for $-1 \leqslant x \leqslant 4$.

x	-1	0	1	2	3	4
$-x^2$	-1	0	-1	-4	-9	-16
$3x$	-3	0	3	6	9	12
$y=-x^2+3x-2$	-6	-2	0	0	-2	-6

In this table you notice that the values of y are symmetrical about the space between the two zeros. Thus, the maximum or minimum point (do you know which?) will lie midway between $x=1$ and $x=2$.

Thus $x=1\frac{1}{2}$,

i.e. $y=-(1\frac{1}{2})^2+3\times(1\frac{1}{2})-2$

$$=-2\tfrac{1}{4}+4\tfrac{1}{2}-2$$

$$=\tfrac{1}{4}$$

Therefore the maximum value of y is $\frac{1}{4}$. See Fig. 3.

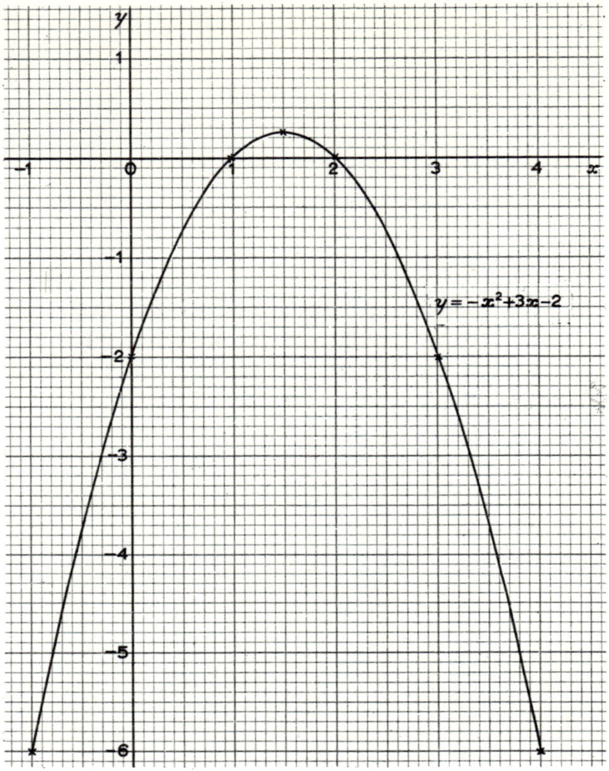

Fig. 3

Exercise 6

1 Draw the graph of $y=x^2-3x-1$ for $-1 \leqslant x \leqslant 4$. What is the minimum value of y? Where does this graph cut the lines $y=0$, $y=-2$, $y=2$?

2 Draw the graph of $y=(x+1)(2-x)$ for $-2 \leqslant x \leqslant 3$. Write down the co-ordinates at the maximum point. Where does this graph cut $y=0$, $y=-x$ and $y=2x+1$?

3 Draw the graph of $\{(x, y): y=x^2-x\}$ for the range $-2 \leqslant x \leqslant 3$. Give the minimum value of y and the co-ordinates of the points where the curve crosses the axes.

4 Find the co-ordinates of the points where $y=2x^2-3x+4$ crosses $y=4$.

5 Find the common points of $\{(x, y): y=(2-x)(x-1)\}$ and $\{(x, y): y=-2\}$.

The following table of values is for the curve $\{(x, y): y = 2x^2 - x + 1\}$ for $-2 \leqslant x \leqslant 3$.

x	-2	-1	0	1	2	3
$2x^2$	8	2	0	2	8	18
$-x$	2	1	0	-1	-2	-3
$y = 2x^2 - x + 1$	11	4	1	2	7	16

These values of y do not have the symmetry we found with previous sets although $y = 2x^2 - x + 1$ is none the less a parabola, see Fig. 4.

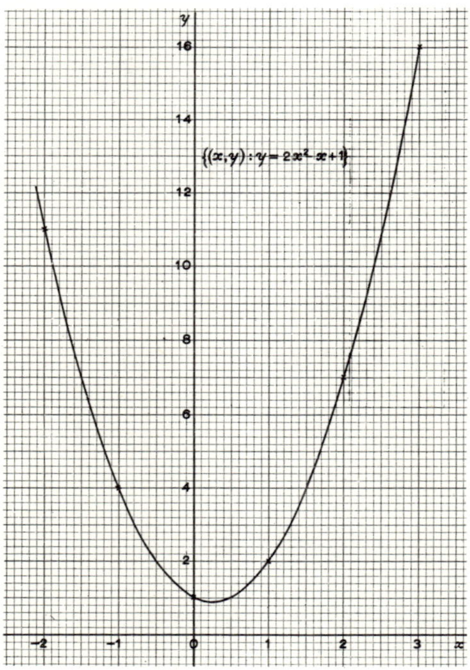

Fig. 4

The values of y decrease from $x=-2$ to $x=0$ and increase from $x=1$ to $x=3$; therefore there will be a minimum point between $x=0$ and $x=1$. The values of y are not symmetrical because the minimum does not lie midway between $x=0$ and $x=1$. Working out the value of y when $x=\frac{1}{2}$ gives the point $(\frac{1}{2}, 1)$ as on the curve.

Thus, both (0, 1) and $(\frac{1}{2}, 1)$ are on the curve. Therefore the minimum point will be given when $x=\frac{1}{4}$.

i.e.
$$\begin{aligned} y &= 2(\tfrac{1}{4})^2-(\tfrac{1}{4})+1 \\ &= \tfrac{1}{8}-\tfrac{1}{4}+1 \\ &= \tfrac{7}{8} \end{aligned}$$

Therefore the co-ordinates of the minimum point are $(\frac{1}{4}, \frac{7}{8})$.

Exercise 7

1 Draw the graph of $y=-2x^2-x+3$ for $-3 \leqslant x \leqslant 2$.

What are the co-ordinates of the maximum point? Where does the curve cross the x-axis?

2 Draw the graph of $\{(x, y): y=(1-x)(2x+1)\}$ for $-2 \leqslant x \leqslant 3$. Give the maximum value of y and the co-ordinates of the points where the line crosses $y=5x+5$.

3 Draw the graph of $y=4x^2+2x-1$ for $-3 \leqslant x \leqslant 2$.

Where does this cross the line $y+6x+1=0$?

4 Find the maximum or minimum values of the following.

Illustrate graphically.

$$y+x^2=x-5 \qquad y+2=x^2+4x$$

5 Where does $y=-x$ cross $y=2x^2+x$?

6 Give the points of intersection of $y=2x^2+3x-4$ and $y=-2x^2-3x+4$.

7 Draw the graph of $\{(x, y): y=3x^2+x-2\}$.

What are the co-ordinates of the minimum point? What is the equation of the axis of symmetry?

8 Find the maximum point of $y=-5x^2+4$.

Solving equations

Drawing graphs is a useful way of solving equations. To solve $x^2-x-2=0$ you would use factors.

i.e. $$x^2-x-2=0$$

$$(x+1)(x-2)=0$$

either $x+1=0$ or $x-2=0$

i.e. $x=-1$ or $x=2$

However, if there are no simple factors another method must be used. This could be completing the square, using the formula or perhaps drawing a graph. Let us solve $x^2-3x-2=0$ graphically.

Method 1

First we need to draw the graph of $y=x^2-3x-2$. Then we will consider where $x^2-3x-2=0$, i.e. where $y=x^2-3x-2$ and $y=0$ cross. Let us choose a range of $-3\leqslant x\leqslant 3$ and hope this will include the solution.

x	-3	-2	-1	0	1	2	3	$1\frac{1}{2}$
x^2	9	4	1	0	1	4	9	$2\frac{1}{4}$
$-3x$	9	6	3	0	-3	-6	-9	$-4\frac{1}{2}$
$y=x^2-3x-2$	16	8	2	-2	-4	-4	-2	$-6\frac{1}{4}$

You will see that it is necessary to work out the value of y when $x=1\frac{1}{2}$ (to obtain the position of the minimum point). When $y=x^2-3x-2$ is drawn, see Fig. 5, you will notice that one of the points where it crosses $y=0$ is shown but the other is not within the range $-3\leqslant x\leqslant 3$. We must therefore extend this range. One can see that the curve will cross $y=0$ again between $x=3$ and $x=4$ so we need only extend as far as $x=4$. $x=4$ gives $y=2$. In Fig. 5 this part of the curve has been added as a dotted line. The solutions to $x^2-3x-2=0$ are given where $y=x^2-3x-2$ cuts $y=0$. That is, when $x=3{\cdot}55$ and $-0{\cdot}55$.

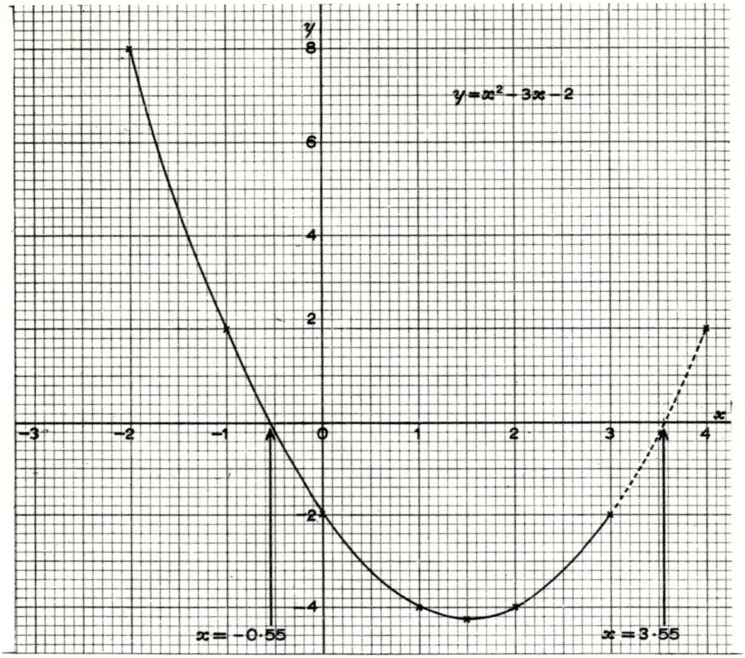

Fig. 5

Exercise 8

Solve the following equations graphically:

1 $x^2-2x-1=0$

2 $x^2+x-3=0$

3 $2x^2+x-4=0$

4 $-x^2-x+3=0$

5 $5+x-3x^2=0$

6 $3x^2+7x+1=0$

7 $4x^2+12x+9=0$

8 $2x^2-5x+3=0$

9 $6+x-x^2=0$

10 $16x^2+8x-3=0$

11 $12-4x-x^2=0$

12 $2x^2-3x+1=0$

Method 2

Working out a table of values and drawing a curve takes quite a long time. It is therefore often quicker to think before embarking on the drawing of a complicated parabola. Let us again consider the equation $x^2-3x-2=0$. This is the same equation as $x^2=3x+2$.

Now $y=x^2$ is a very simple parabola (you can probably work out the necessary values of y in your head) and $y=3x+2$ is a straight line. Thus if we draw $y=x^2$ and $y=3x+2$, the solutions we require will be given where the graphs intersect.

Fig. 6

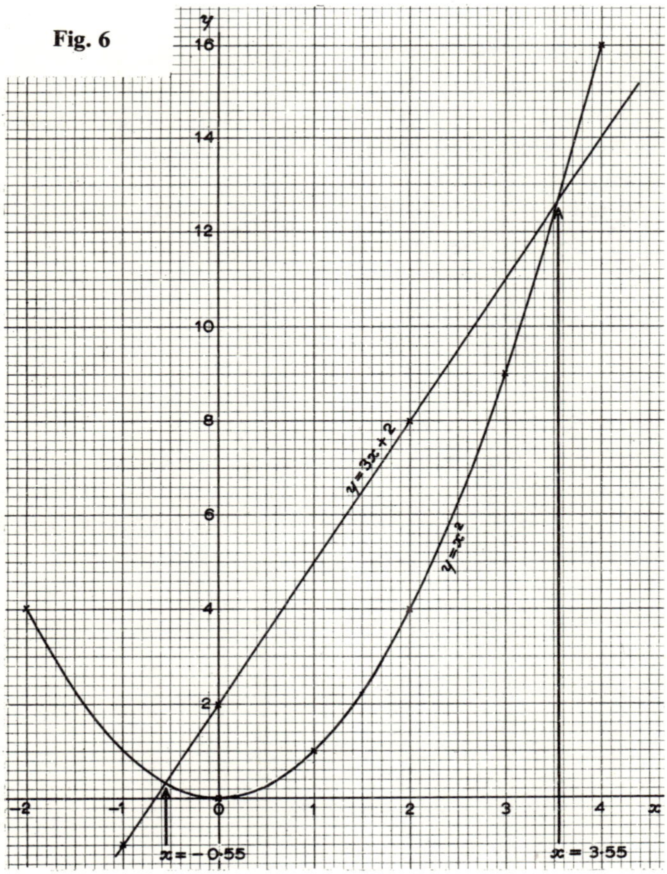

Exercise 9

1 Solve again the equations given in Exercise 8 using the method indicated on page 23. (You can use the same graph of $y=x^2$ more than once.)

2 Solve the following quadratic equations by separating the term in x^2 from the linear part and drawing graphs of the two parts:

(a) $-x^2+x+4=0$
(b) $5-2x-x^2=0$
(c) $2x^2-4x+1=0$
(d) $2x^2-x-7=0$
(e) $-3x^2+4x+1=0$
(f) $2-x-3x^2=0$

The wording of the question will decide which of the methods indicated is most suitable. Method 2 is quicker if only the solutions to equations are required; but if other information is required about the parabola, then Method 1 (or a combination) may be better. Consider the following worked example.

Find the maximum point of $y+x^2=2x+2$. Where does this parabola cut the axes? Solve the equations:

$$-x^2+2x+2=0$$
$$2x+1=x^2$$
$$4x-1=x^2$$

To find the maximum point and answer the part of the question about cutting the axes, it is necessary to draw $y=-x^2+2x+2$. Once this has been drawn it is quicker to use it to solve the equations than to start on Method 2.

x	-3	-2	-1	0	1	2	3	4
$-x^2$	-9	-4	-1	0	-1	-4	-9	-16
$2x$	-6	-4	-2	0	2	4	6	8
$y=-x^2+2x+2$	-13	-6	-1	2	3	2	-1	-6

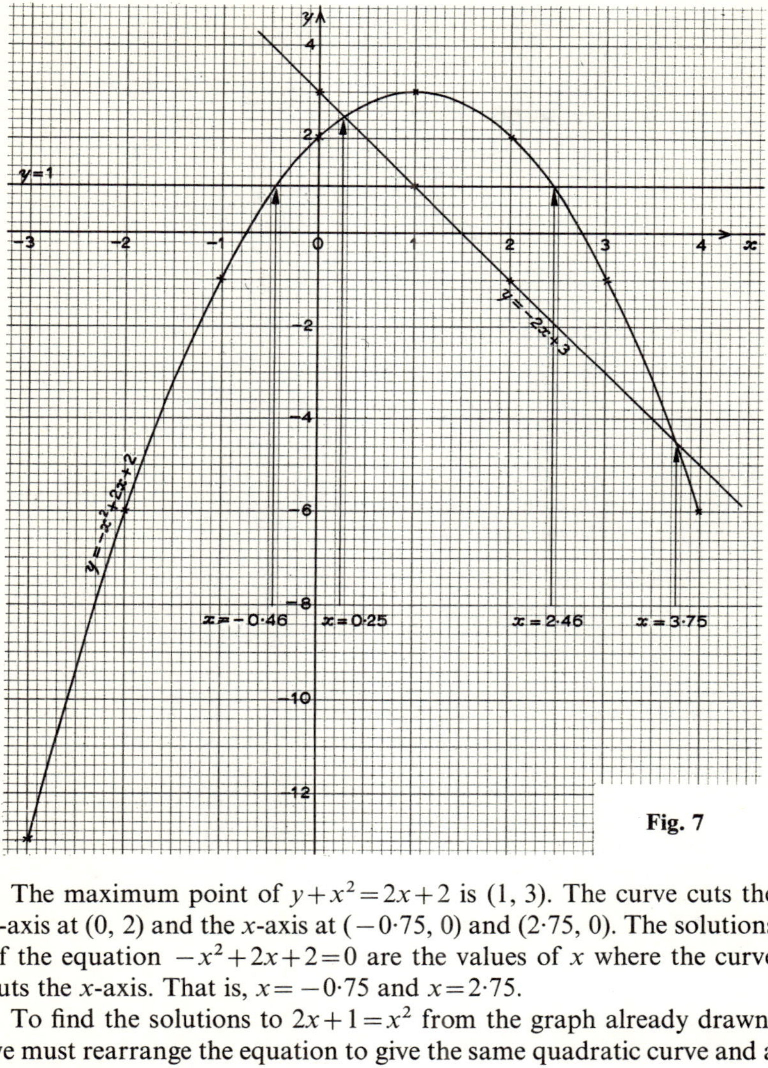

Fig. 7

The maximum point of $y+x^2=2x+2$ is (1, 3). The curve cuts the y-axis at (0, 2) and the x-axis at $(-0{\cdot}75, 0)$ and $(2{\cdot}75, 0)$. The solutions of the equation $-x^2+2x+2=0$ are the values of x where the curve cuts the x-axis. That is, $x=-0{\cdot}75$ and $x=2{\cdot}75$.

To find the solutions to $2x+1=x^2$ from the graph already drawn, we must rearrange the equation to give the same quadratic curve and a straight line.

i.e.
$$2x+1=x^2$$
$$-x^2+2x+1=0$$
$$-x^2+2x+2=1$$

Therefore the solutions are given where $y=-x^2+2x+2$ cuts $y=1$.

i.e. $x=-0{\cdot}46$ and $x=2{\cdot}46$

Similarly for the solutions of $4x-1=x^2$. Rearrangement into $-x^2+2x+1=$ a linear function.

i.e.

$$4x-1=x^2$$
$$-x^2+4x-1=0$$
$$-x^2+2x-1=-2x$$
$$-x^2+2x+2=-2x+3$$

Therefore the solutions are given where $y=-x^2+2x+2$ cuts $y=-2x+3$.

i.e. $x=0{\cdot}25$ and $x=3{\cdot}75$

Exercise 10

1 Find the minimum point of $y=2x^2-x-1$. Does $y=2x^2-x-1$ cut the following lines? If so, where?

(a) $y=0$ (b) $y=-4$ (c) $y=2$

Find the solutions to the following equations:

$$2x^2-x-1=0$$
$$2x^2=x+1$$
$$2x^2=x$$
$$2x^2-x=-1$$

2 State where $y=x^2-2x+1$ cuts the co-ordinate axes. Does it have a maximum or a minimum point? Give the value of y at this point.

Solve:

$$(x-1)^2=0$$
$$(x-1)(x-2)=0$$
$$x^2=2x+3$$

3 Solve: $2x^2+3x-1=0$ and $2x^2+x-5=0$.

4 Solve: $4x+1=x^2$ and $3x-x^2=1$.

5 Draw the graph of $\{x, y: y=5-x^2\}$. Use this to find $\sqrt{5}$ and to solve $x^2-x=5$.

6 Draw the graph of $y=x^2$. Use this to find $\sqrt{3}$, $\sqrt{8}$ and $\sqrt{10}$. Solve the following equations without drawing any other parabola:

$$x^2=x+1$$
$$x^2-2=0$$
$$3x=2-x^2$$

7 Find the maximum point of $2x+1=y+3x^2$. Where does this parabola cross the line $y=x$? The x-co-ordinates at the points of intersection of $y=x$ with $2x+1=y+3x^2$ are the solutions to an equation. State this equation.

8 State the equation whose solutions are given by the intersection of $y=x^2+5x-2$ and $y=1$. Find the solutions to this equation by the quickest method.

9 State the equation whose solutions are given by the intersection of $\{(x, y): y=-2x^2+x+1\}$ and $\{(x, y): y=x-6\}$.
Find the solutions to this equation by the quickest method.

10 Solve:

$$x^2+4x-3=0, \quad x^2+4x-2=0, \quad x^2+4x-1=0, \quad x^2+4x=0,$$
$$x^2+4x+1=0, \quad x^2+4x+2=0, \quad x^2+4x+3=0.$$

Some quadratic equations do not have solutions. For example there are no values of x which would solve $x^2+x+4=0$, or $x^2=-5$. Fig. 8 shows the difference between equations which have solutions and equations which do not.

Fig. 8(a) shows $y=x^2+2x-2$. This parabola cuts $y=0$ twice, and thus $x^2+2x-2=0$ has two solutions, i.e. $x=-2{\cdot}75$ and $x=0{\cdot}75$.

Fig. 8(b) shows $y=x^2+2x+3$. This parabola does not cut $y=0$. Thus $x^2+2x+3=0$ does not have any solutions.

Fig. 8(c) shows the borderline between the two cases. This is the graph of $y=x^2+2x+1$ or $y=(x+1)^2$.

We say that this has two equal solutions $x=-1$.

Think carefully about the positions of these parabolas and their equations. Can you see any connection?

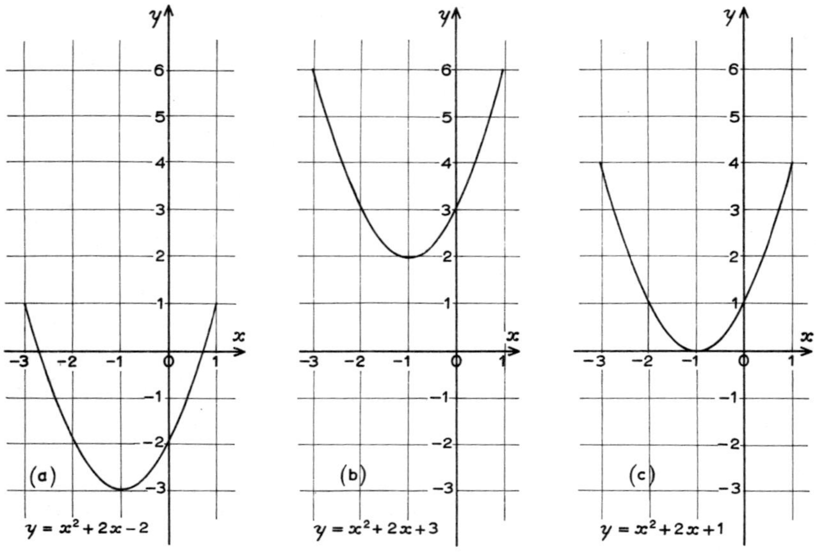

Fig. 8

Exercise 11

Draw graphs to determine which of the following equations have solutions.

1 $x^2+x+4=0$
$x^2+x-2=0$
$x^2+x-1=0$
$x^2+x+1=0$

2 $x^2-4x+6=0$
$x^2-4x+4=0$
$x^2-4x-2=0$
$x^2-4x+1=0$

3 $5-2x-x^2=0$
$-3-2x-x^2=0$
$-1-2x-x^2=0$
$2-2x-x^2=0$

4 Consider what would happen if you tried to answer the previous three questions using Method 2. Illustrate your answer to at least one of these questions.

Revision Exercise I

1 Draw $3y+4=x$ for $-4 \leqslant x \leqslant 4$. What is the gradient of this line and where does it cross the axes?

2 Give the equations of the lines joining the following pairs of points:

(a) $(0, 5):(3, 1)$ (b) $(4, 2):(-2, -1)$
(c) $(-3, 1):(2, 1)$ (d) $(-4, -3):(1, 2)$
(e) $(-2, -2):(0, 2)$

3 Temperature falls by approximately 1°C for each increase in height of 180 metres. If the temperature at sea level is 15°C, draw a graph to determine the relationship between temperature, θ, and height, h. Read from your graph the temperatures at 300 metres and 600 metres above sea level.

4 Draw the graph of $y=x^2-5x+2$ for $-1 \leqslant x \leqslant 5$. Give the equation of the axis of symmetry and the co-ordinates of the minimum point. Where does this parabola cross the lines $y=4$, $y=0$, $y=-4$?

5 Draw the graphs of $y=2x^2$ and $y=3x-1$. The x-co-ordinates of the points where these graphs cross are the solutions to an equation. State this equation and give its solutions.

6 Solve the following equations graphically:

$$y=-x^2+3x+2$$
$$y=-x^2-2x+1$$
$$y=-x^2-2$$

7 A car covers a certain distance after the brakes have been applied before it stops. The following are stopping distances d for cars travelling with speeds s.

Speed in km/h, s	20	30	40	50
Stopping distance in m, d	15	40	68	100

Draw a graph to determine the relationship between s and d.

How far would the car travel (after the brakes were applied) if it were initially travelling at 45 km/h?

Chapter Three

INEQUALITIES

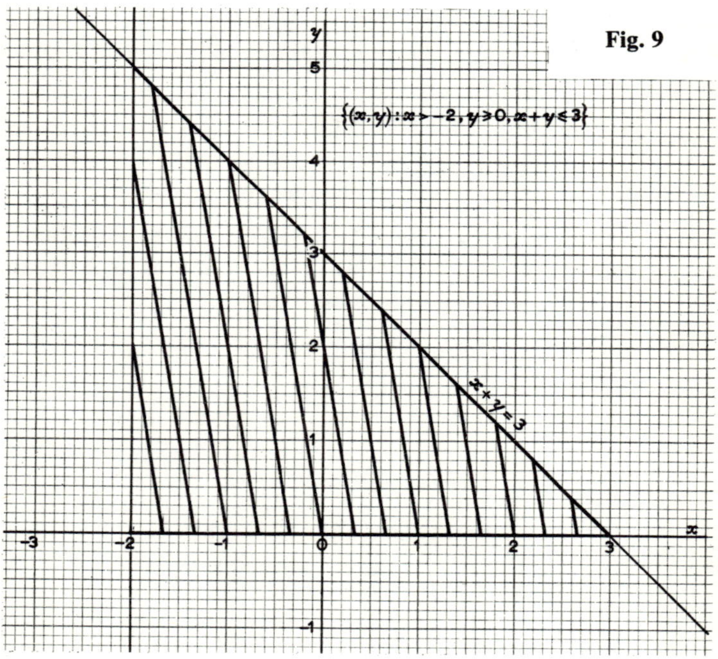

Fig. 9

Exercise 12

Consider Fig. 9 carefully and then illustrate the following sets. Use coloured pencils for clarity and make sure your diagrams show the difference between $\geqslant$ and $>$.

1 $\{(x, y): y \geqslant 2\}$ **2** $\{(x, y): x \leqslant -1\}$

3 $\{(x, y): x < -1\}$ **4** $\{(x, y): y > x\}$

5 $\{(x, y): y \geqslant 2x + 1\}$ **6** $\{(x, y): y < -3x + 2\}$

7 $\{(x, y): y \leqslant x + \frac{1}{2}\}$

8 $\{(x, y): x < 1,\ y \geqslant 0\}$

9 $\{(x, y): x > 2,\ y \leqslant -3\}$

10 $\{(x, y): x \geqslant 5,\ y \leqslant x\}$

11 $\{(x, y): y \geqslant 5 - x$ and $y < x\}$

12 $\{(x, y): x \geqslant 0,\ y \geqslant 0$ and $y < x + 5\}$

13 $\{(x, y): x \geqslant 0,\ y \geqslant 0$ and $x + y \leqslant 6\}$

14 $\{(x, y): x \geqslant 1,\ y \geqslant 0$ and $2x + y \leqslant 10\}$

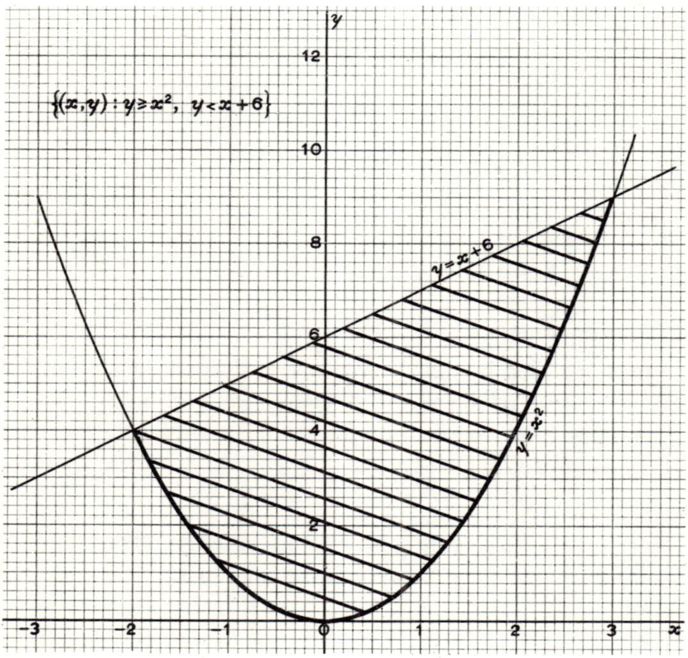

Fig. 10

Fig. 10 illustrates $\{(x, y): y \geqslant x^2$ and $y < x + 6\}$.

Notice that $y > x^2$ is the part of the page above the line $y = x^2$.

It is easy to check which side of the curve is required by considering any one point.

Example
Consider the point (0, 2).

$$y=2$$
$$x^2=0$$
$$\therefore\ y>x^2$$

Thus (0, 2) lies in the half-plane $y>x^2$ and also $y>x^2$ is the part of the page bounded by $y=x^2$ which contains (0, 2).

Exercise 13
Illustrate the following:

1 $\{(x, y): y \geqslant x^2, y \leqslant 2x+1\}$
2 $\{(x, y): y>(x+1)^2, y<6\}$
3 $\{(x, y): y>(x+1)^2, y \leqslant 2x+1\}$
4 $\{(x, y): y \leqslant 5-x^2, y>0\}$
5 $\{(x, y): y \leqslant 2+x-x^2, y<-2\}$
6 $\{(x, y): y \leqslant x^2, y>-1\}$
7 $\{(x, y): y \leqslant x^2, y<4\}$
8 $\{(x, y): y>x^2+3x+1, y>x+1\}$
9 $\{(x, y): x \geqslant 0, y \geqslant 0, y \geqslant x^2-2\}$
10 $\{(x, y): x<6, y \geqslant 0, y<2x^2+4\}$

Linear programming

Inequalities are used by many large organisations such as industry or the army to solve problems of efficiency and economy. Many such problems require the use of computers, but in simple cases it is possible to find the answer graphically.

Example
A market gardener is deciding how to allocate the 4000 square metres of greenhouse space. He grows tomatoes and cucumbers and estimates as follows.

	tomatoes	cucumbers
Cost per 100 square metres in £	2	1
Labour time per 100 square metres in hr	3	2
Profit per 100 square metres in £	4	3

The maximum cost must not exceed £60 and the maximum labour time available will be 96 hours. How is he to allocate the space to gain the maximum profit?

Let x be the number of units of 100 square metres planted with tomatoes and y be the number of units of 100 square metres planted with cucumbers.

Then since the cost is limited to £60 $\quad 2x+y \leqslant 60$

Since the labour time must not exceed 96 hours $\quad 3x+2y \leqslant 96$

Also it is not possible to plant a negative amount $\quad x \geqslant 0$

$y \geqslant 0$

Only 4000 square metres are available therefore $\quad x+y \leqslant 40$

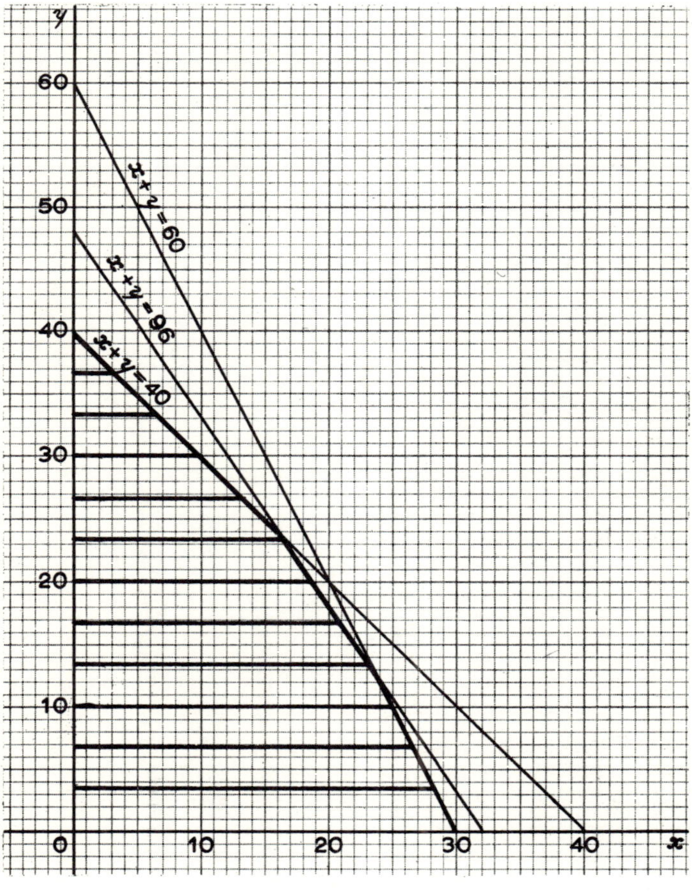

Fig. 11

In Fig. 11 the shaded area is the section of the page where all the inequalities are true at the same time. Therefore the solution to the problem of allocation of space will be given by any point within this area. However, to ensure maximum profit we must go further.

The profit will be £$(4x+3y)$.

Let us estimate the profit as £60, i.e. $4x+3y=60$.

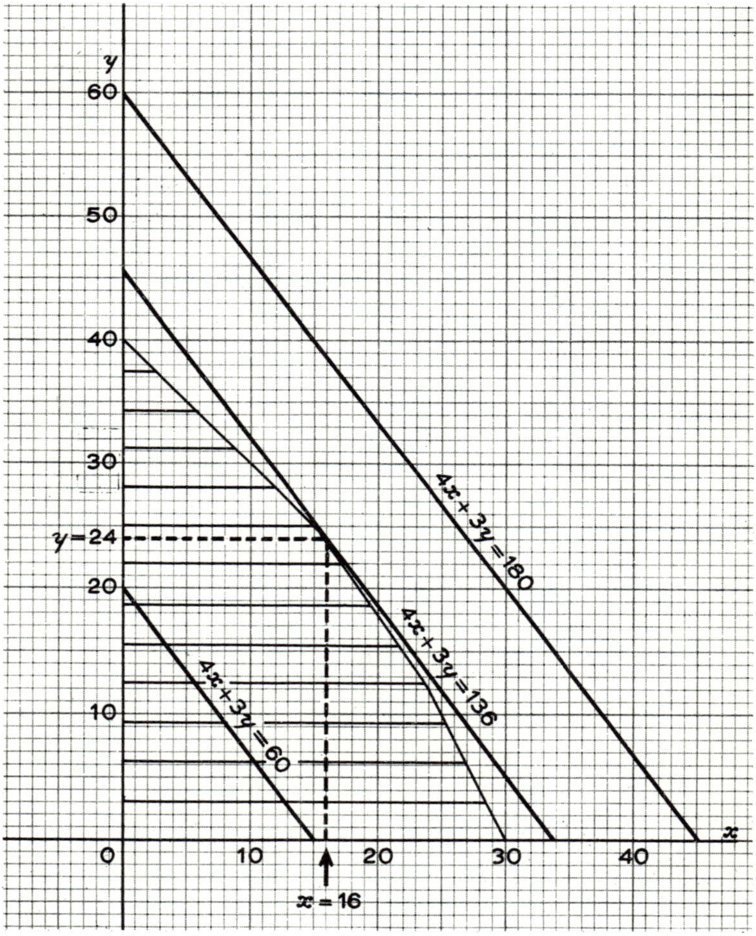

Fig. 12

Fig. 12 comprises the same shaded area as Fig. 11 with the addition of various parallel lines. One of these lines is $4x+3y=60$. You will see that there are many points of the line $4x+3y=60$ which lie in the shaded area. There are therefore many ways of planting to gain £60 profit.

Let us estimate the profit to be £180, i.e. $4x+3y=180$.

You will see that no point of $4x+3y=180$ lies within the shaded area. It is therefore impossible to make a profit of £180.

All 'profit lines' will be parallel as all have the form $4x+3y=$constant. The larger the value of the constant, the larger the profit—provided the line contains at least one point of the shaded area. You will see that 136 is the largest possible value for the constant if the line $4x+3y=$constant is to pass through a point of the shaded area. The point which $4x+3y=136$ has in common with the shaded area is (16, 24). Thus by planting 1600 square metres of tomatoes and 2400 square metres of cucumbers the maximum profit of £136 is obtainable. This process for finding the maximum profit (or minimum cost) is called maximising (or minimising).

Exercise 14

1 A supermarket sells two brands of instant coffee, X and Y. At least twice as much of brand X is sold as brand Y. There is 20p profit on a box of brand X but 30p profit on a box of brand Y. If the available storage space will hold up to 90 boxes, how many boxes of each kind should be stored to give the maximum profit? What will this profit be?

2 Fog causes aircraft to land at an unscheduled airfield. A coach operator is approached to provide transport for 560 people and 15 tons of luggage. He has two types of coach available in sufficient numbers. Type A can carry 40 passengers and 1 ton of luggage. Type B can carry 35 passengers and 1 ton of luggage. The operating costs for the journey will be Type A, £12, and Type B, £10. How many of each sort of coach should be used for the cost to be least? What is the minimum cost?

3 A breeder can buy two different foods for his horses. Each horse needs 6 lb of protein and 1 lb of minerals per week.

	Protein per lb	Mineral per lb	Cost per lb
Feed *A*	12 oz	4 oz	$62\frac{1}{2}$p
Feed *B*	15 oz	1 oz	50p

How should the breeder mix the feeds to gain the correct requirements at minimum cost? What is the breeder's weekly bill if he keeps 12 horses?

4 A firm weaves two kinds of cloth from terylene and wool. The first kind contains 50% of each fibre and the second 80% terylene to 20% wool. Type 1 sells at a profit of £2 per roll while Type 2 makes a profit of £2·50 per roll. There is sufficient fibre to weave 90 rolls, $\frac{2}{3}$ of this being terylene. What is the maximum profit?

Chapter Four

OTHER CURVES

The only curves so far encountered in this book have been parabolas (readers of *Graphs and Co-ordinates* in Series I will have met one or two other sorts). A parabola is the shape determined by an equation of the type $y=ax^2+bx+c$ where a, b and c are constants. There are many other types of curves and we shall consider some of these now.

The cubic

If we introduce a term in x^3, we have the equation of a cubic. For example, $y=ax^3+bx^2+cx+d$ where a, b, c and d are constants.

Let us first consider the simplest form $y=x^3$ for $-3 \leqslant x \leqslant 3$.

x	-3	-2	-1	0	1	2	3
$y=x^3$	-27	-8	-1	0	1	8	27

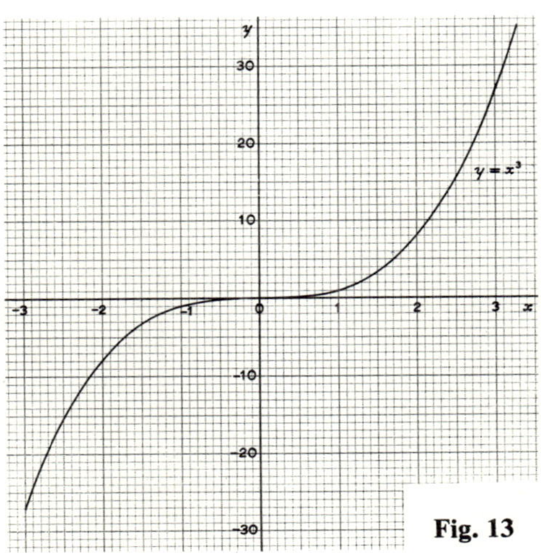

Fig. 13

Exercise 15

1 Draw the following from $-3 \leqslant x \leqslant 3$ on the same sheet of paper:

$y=x^3 \quad y=2x^3 \quad y=3x^3$

2 Draw the following (in the range $-3 \leqslant x \leqslant 3$) on the same sheet of paper:

$y=x^3 \quad y=-x^3$

3 Draw $y^3=x$ for $-4 \leqslant y \leqslant 4$. Use your graph to find:

$\sqrt[3]{50} \quad \sqrt[3]{20} \quad \sqrt[3]{35}$.

4 On the same sheet of graph paper draw

$$y=x^3$$
$$y=x^3+1$$
$$y=x^3+2$$
$$y=x^3-1$$

for the range $-3 \leqslant x \leqslant 3$.

We will now consider a more general cubic.

$$y=x^3+2x^2-x-2 \qquad -3 \leqslant x \leqslant 2$$

x	-3	-2	-1	0	1	2
x^3 $2x^2$ $-x$	-27 18 3	-8 8 2	-1 2 1	0 0 0	1 2 -1	8 8 -2
y	-6	0	0	-2	0	12

The maximum and minimum points of the cubic in Fig. 14 are not at integral values of x and y. The positioning of the maximum and minimum points on the curve must be done by trial and error at this stage. You must place them so that the resulting curve looks reasonable.

You will notice that $y=x^3+2x^2-x-2$ cuts the x axis in three places. These are the solutions of $x^3+2x^2-x-2=0$. As they are -2, -1 and $+1$, they are also the solutions of $(x+2)(x+1)(x-1)=0$ since $(x+2)(x+1)(x-1)$ is zero if any of its factors are zero. Check, by expanding the brackets, that $y=(x+2)(x+1)(x-1)$ is in fact the same equation as $y=x^3+2x^2-x-2$.

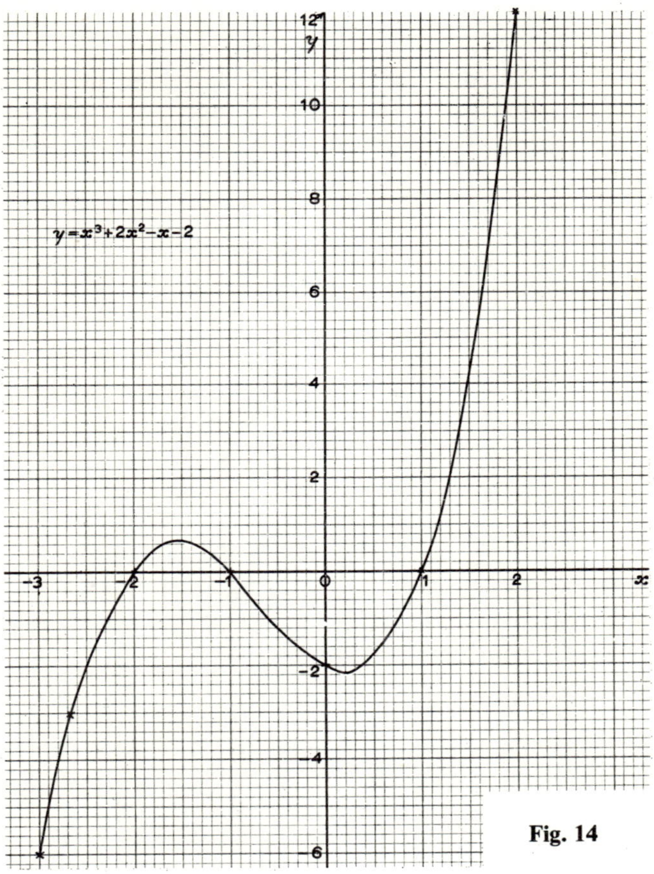

Fig. 14

Exercise 16

1 Draw the graph of $y=x^3-x$ for $-3\leqslant x\leqslant 3$. Solve the equations

$$x^3-x=0$$

and $$x^3-x=2$$

2 Draw $y=x^3-2x+1$ for $-2\leqslant x\leqslant 3$. Solve the equations

$$x^3-2x+1=0$$

and $$x^3-2x=0$$

Give the positions of the maximum and minimum points of $y=x^3-2x+1$.

3 Draw $y=2x-x^2-x^3$ for $-3 \leqslant x \leqslant 2$. Give the maximum and minimum values of $y=2x-x^2-x^3$, and solve the equation $x^3+x^2=2x$. Give only the answers to these questions which lie in the range $-3 \leqslant x \leqslant 2$.

4 Draw $y=x^2(1+x)$ for $-2 \leqslant x \leqslant 2$. How many times does this cut the following lines?

$y=2, \quad y=\frac{1}{10}, \quad y=-4, \quad y=x.$

5 Draw $y=x^3-2x^2+x+1$ for $-2 \leqslant x \leqslant 3$. Solve the equations

$$x^3-2x^2+x+1=0$$

and

$$x^3-2x^2+x=0$$

6 Draw the graph of $y=x(x+1)(2x-1)$ for $-2 \leqslant x \leqslant 2$.
Give the maximum and minimum values of $y=2x^3+x^2-x$ and solve the equation $2x^3+x^2=x$.

7 Draw the graphs of $y=x(x^2-1)$ and $y=1-x^2$ for $-3 \leqslant x \leqslant 3$.
Solve the following equations:

$$x(x^2-1)=6$$
$$1-x^2=0$$
$$x^3+x^2-x-1=0$$

For the solutions of a cubic equation it is often more straightforward to split the equation into two parts and draw two graphs.

Consider again the cubic $y=x^3+2x^2-x-2$. To solve $x^3+2x^2-x-2=0$, we rewrite the equation as $x^3=-2x^2+x+2$ and then draw $y=x^3$ and $y=-2x^2+x+2$. Where these curves cross, give the solutions to the original equation.

i.e. $x=-2, -1$ and 1. See Fig. 15 on page 41.

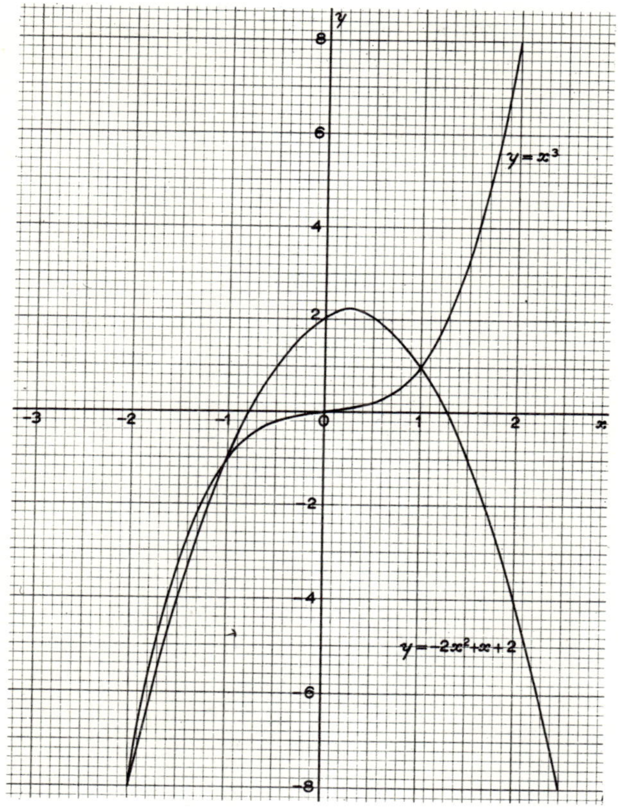

Fig. 15

Exercise 17

Solve the following equations by drawing two curves in the simplest possible cubic curve and a quadratic or lesser order curve.

1 $x^3-4x=0$

2 $x^3+6x^2+11x+6=0$

3 $2x^3-x^2-2x+1=0$

4 $2+x-2x^2-x^3=0$

5 $x^3+x-1=0$

The hyperbola

$y=1/x$ is a simple hyperbola. Let us consider it first in the range $1 \leqslant x \leqslant 5$.

x	1	2	3	4	5
$y=1/x$	1	0·5	0·33	0·25	0·20

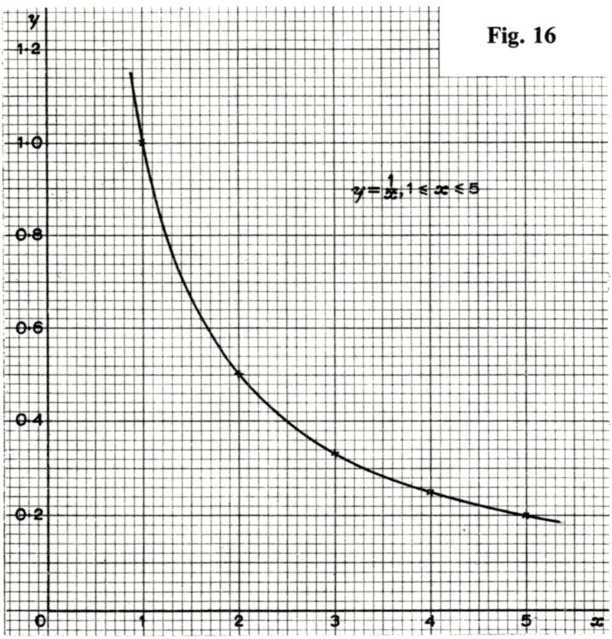

Fig. 16

You will see this illustrated in Fig. 16. Notice that a large scale is required on the y-axis. Let us now consider what happens if we extend our range.

$$x=10 \quad \Rightarrow y=0{\cdot}10$$
$$x=50 \quad \Rightarrow y=0{\cdot}02$$
$$x=100 \Rightarrow y=0{\cdot}01$$
$$x=1000 \Rightarrow y=0{\cdot}001$$

As x increases y decreases but always remains positive. The x-axis and the curve gradually approach closer and closer but never meet.

$$x=\tfrac{1}{2} \Rightarrow y=2$$
$$x=\tfrac{1}{10} \Rightarrow y=10$$
$$x=\tfrac{1}{50} \Rightarrow y=50$$
$$x=\tfrac{1}{100} \Rightarrow y=100$$

As x becomes closer to zero y increases to large values.

The y-axis and the curve gradually approach closer and closer but never meet.

A similar situation occurs for negative values of x. Verify this for yourself and compare your results with Fig. 17 which illustrates the hyperbola $y=1/x$ for $-10 \leqslant x \leqslant 10$.

Notice that the hyperbola is a discontinuous curve.

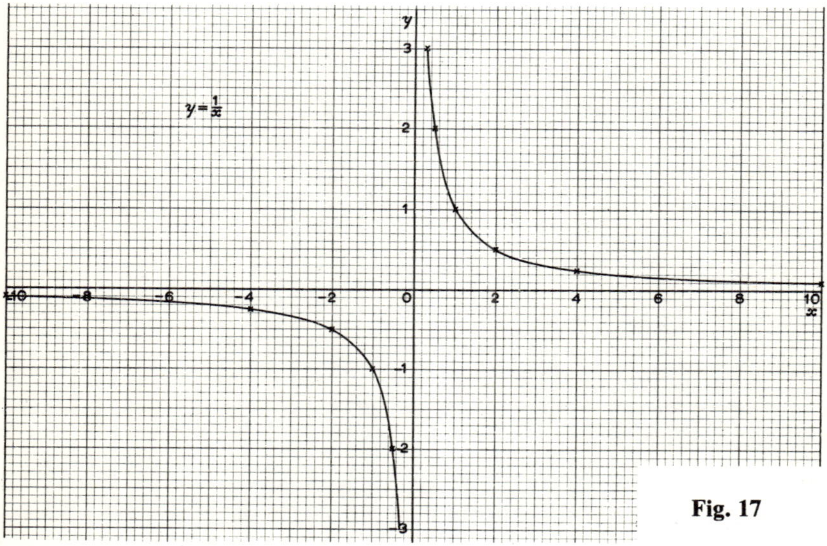

Fig. 17

Exercise 18

1 Draw the graph of $y=\dfrac{1}{x}+1$ for $-5 \leqslant x \leqslant 5$. Solve the equations

$$\frac{1}{x}+1=0 \quad \text{and} \quad \frac{1}{x}+1=x^2$$

2 Draw the graph of $y=\frac{2+x}{x}$ for $1\leqslant x\leqslant 5$. Solve the equation

$$\frac{2+x}{x}=2\cdot 5$$

3 Draw the graph of $xy=2$ for $-5\leqslant x\leqslant 5$. Solve the equation

$$\frac{2}{x}=x$$

4 Draw the graph of $y=1/x^2$ for $1\leqslant x\leqslant 4$.

Consider carefully what happens when the range of values of x is extended, and then draw the graph of $y=1/x^2$ for $-5\leqslant x\leqslant 5$.

5 Draw on the same page the graphs of:

$$y=\frac{1}{x^2}$$

$$y=\frac{1}{x^2}+1$$

$$y=\frac{1}{x^2}+2$$

$$y=\frac{1}{x^2}-1$$

$$y=\frac{1}{x^2}-2$$

for the range $-5\leqslant x\leqslant 5$.

6 Draw on the same axes the graphs of $xy=1$, $xy=4$, $xy=10$. Do you notice anything about these curves? Where are their axes of symmetry?

7 An aeroplane travels at 500 m.p.h., a sports car at 100 m.p.h., a lorry at 40 m.p.h., a cyclist at 12 m.p.h. and a man at 4 m.p.h. Draw a graph to illustrate the times taken to cover a journey of 100 miles.

Sine curve

Consider a line of unit length revolving in an anti-clockwise circle. Assume the line started horizontal to the right of the pivot point as illustrated in Fig. 18.

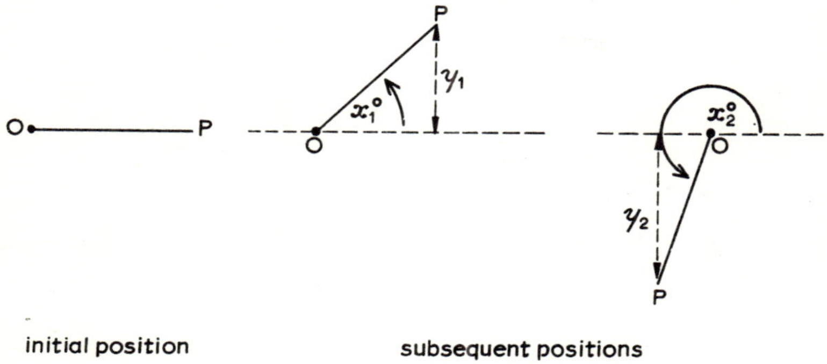

Fig. 18

We are going to consider how the distance y of the point P from the initial line varies as the angle x changes. In Fig. 18 you will see two possible positions. Notice that y_2 is a negative measurement as P is below the initial line in this case.

Example

On a sheet of graph paper, take 2 inches to represent 1 unit and draw a circle of unit radius. Now let the point P travel anti-clockwise round the circle and mark its position on the circumference for every change of 15° at the centre of the circle. In each case measure the distance of P from the initial line and write out a table of values as below: do not forget that your unit of measurement is 2 inches.

x°	0	15	30	45, etc.,	360
Distance of P from initial line, y	0		0·5		

Now plot y against x and draw a curve. Can you extend this beyond 360°?

Can you extend this curve to include negative angles, i.e. by turning clockwise from the initial position? You should have drawn part of the sine curve which is shown in Fig. 19. This is an example of a periodic function, i.e. a function which regularly repeats its values.

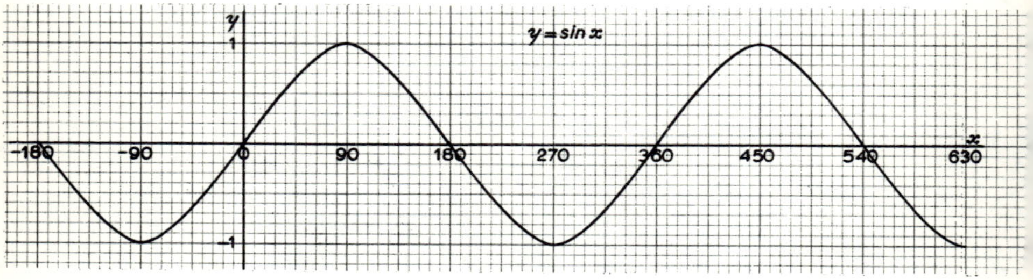

Fig. 19

Exercise 19

Note. A knowledge of radians is required for this exercise.

1 Redraw the graph of $y = \sin x$ for $-2\pi \leqslant x \leqslant 2\pi$. On the same axes draw the graph of $y = x$ and hence solve the equation $x = \sin x$.

2 Draw the graph of $y = \sin 2x$ for $-2\pi \leqslant x \leqslant 2\pi$. Use this to solve the following equations:

(a) $x = \sin 2x$
(b) $x/2 = \sin 2x$

Log and exponential

Given below is a table of values for $y = 5^x$, $0 \leqslant x \leqslant 4$.

x	0	1	2	3	4
y	1	5	25	125	625

Fig. 20 illustrates this curve.

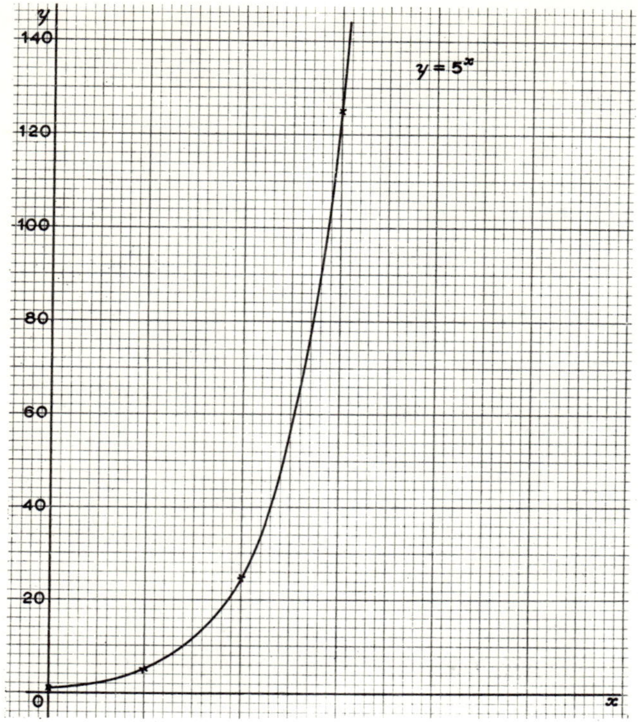

Fig. 20

Exercise 20

1 Work out a table of values for $y=2^x$, $0 \leqslant x \leqslant 5$ and draw the graph of $\{(x, y): y=2^x\}$ for this range. Read off from your graph the values of $2^{1\cdot5}$, $2^{2\cdot5}$, $2^{3\cdot5}$ and $2^{4\cdot5}$.

2 Draw on the same axes the graphs of $y=1^x$, $y=2^x$, $y=3^x$, $y=4^x$, $y=5^x$.

3 Use your normal logarithm tables to work out a table of values for $y=10^x$, $0 \leqslant x \leqslant 1\cdot6$ taking values of x every $0\cdot2$. Use this to solve the equations:

$$x+10=10^x$$

$$2x+4=10^x$$

$$20x=10^x$$

4 Use your table of Naperian logs to work out a table of values for $y=e^x$, $0 \leqslant x \leqslant 3$. Illustrate.

Do you know of any physical applications of e^x?

Revision Exercise II

1 Illustrate the following:

(a) $\{(x, y): y \geqslant x^2, y \leqslant x+1\}$
(b) $\{(x, y): y > 2, x \leqslant -1\}$
(c) $\{(x, y): x \geqslant 0, y \geqslant 2, x+y < 5\}$
(d) $\{(x, y): y > -x^2+4x+1, y < 6\}$

2 A manufacturer makes two types of chocolate. His machines can be easily adapted to make either type. To make a batch of Type *A* requires 2 machines with 3 workmen and shows a profit of £20 per machine per week. To make a batch of Type *B* requires 4 machines with 3 workmen and shows a profit of £30 per machine per week. If there are 30 machines and 24 men available, what is the maximum profit?

3 Give the equations of the straight lines joining the following pairs of points.

(a) (0, 1) (6, 3)
(b) (−2, −8) (4, −2)
(c) (4, 3) (4, −2)

4 Find the solutions to $2x^2-4x+1=0$ and $x^2=3x-5$.

5 Draw the graph of $y=x^3-6x^2+11x-6$ for $-1 \leqslant x \leqslant 4$. Give the positions of the maximum and minimum values and solve $x^3-6x^2+11x=6$.

6 Draw $y=\dfrac{2}{x^2}+1$ for $-4 \leqslant x \leqslant 4$.

Does $y=\dfrac{2}{x^2}+1$ have maximum or minimum points?

Solve $\dfrac{2}{x^2}+1=x$.

Chapter Five

CURVE SKETCHING

Exercise 21

Sketch the graphs of the following curves. Do not use graph paper for this exercise.

1 $y=x$ **2** $y=2$
3 $y=2x+3$ **4** $x=-2$
5 $y=-3x+1$ **6** $y=-\frac{1}{2}x-2$
7 $y=x^2$ **8** $y=x^3$
9 $y=1/x$ **10** $y=1/x^2$

In *Graphs and Co-ordinates* in Series I you did a lot of work on sketching straight lines. We are now going to extend this to some simple curves. Fig. 21 shows sketches of $y=x^2$, $y=x^2+1$, $y=x^2+2$, $y=x^2+3$.

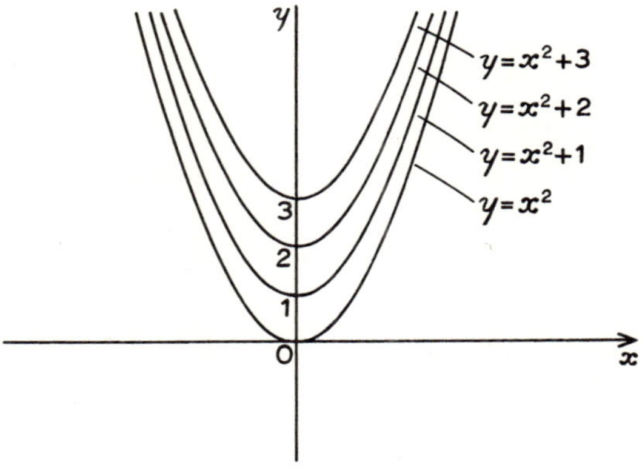

Fig. 21

Using Fig. 21 as a guide, can you sketch $y=x^2$, $y=x^2-1$, $y=x^2-2$, $y=x^2-3$? Notice that the curves do not remain 'parallel' but become closer together at the extremities.

Exercise 22

1 Sketch $y=x^3$, $y=x^3-1$, $y=x^3+1$, $y=x^3+2$.

2 Sketch $y=\frac{1}{x}$, $y=\frac{1}{x}+1$, $y=\frac{1}{x}+2$.

3 Sketch $y=x$, $y=x+2$, $y=x+5$, $y=x-6$.

4 Sketch $y=x^2$, $y=2x^2$, $y=3x^2$, $y=10x^2$. Notice that all these curves pass through the origin.

5 Sketch $y=x^3$, $y=2x^3$, $y=4x^3$, $y=10x^3$.

6 Sketch $y=1/x^2$, $y=2/x^2$, $y=10/x^2$.

7 Sketch $y=x$ and $y=-x$.

8 Sketch $y=x^2$ and $y=-x^2$.

9 Sketch $y=4x^3$ and $y=-4x^3$.

10 Sketch $y=5/x$ and $y=-5/x$.

11 Sketch $y=\sin x$ and $y=-\sin x$, for $0\leqslant x\leqslant 360°$.

If the equation of the parabola or cubic curve is given in factor form the sketching of the curve is made very easy.

Example

(i) $y=(x-1)(x+2)$.

This has a positive term in x^2 and is therefore vertex down.

When $y=0$, $(x-1)(x+2)=0$

i.e. $$x=1 \quad \text{or} \quad x=-2$$

Thus the curve cuts the y-axis at $x=1$ and $x=-2$.

The intercept of the curve with the y-axis is obtained by substituting $x=0$ into the equation,

i.e. $$y=(0-1)(0+2)=-2$$

The axis of symmetry is midway between the intercepts on the x-axis.

i.e. $$x=-\tfrac{1}{2}$$

See Fig. 22(a).

(ii) $y=x(x-1)(x+1)$.

This has a positive term in x^3, and thus the basic line follows the graph of $y=x^3$.

When $y=0$, $x(x-1)(x+1)=0$

i.e. $$x=0 \quad \text{or} \quad x=\pm 1.$$

Thus the curve cuts the x-axis at $x=-1$, $x=0$, $x=1$.

The intercept on the y-axis is at $y=0$.

See Fig. 22(b).

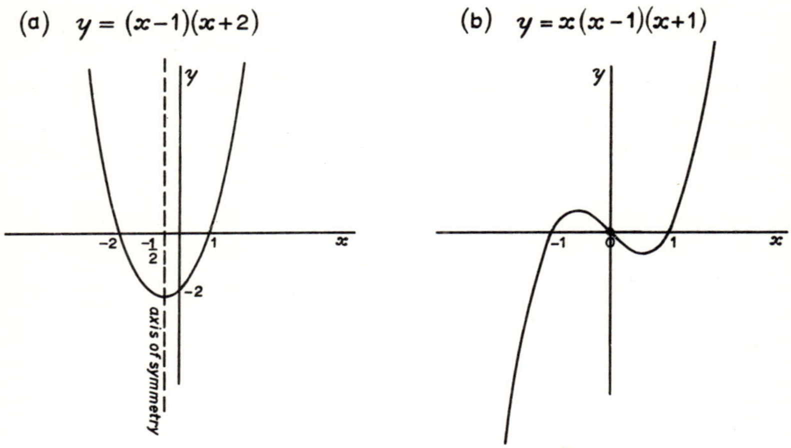

Fig. 22

Exercise 23

Sketch the following curves, giving numerical values for their intercepts on both x and y axes.

1 $y=x(x-1)$
2 $y=(x+1)(2-x)$
3 $y=(3-x)(3+x)$
4 $y=x(x+1)(x+2)$
5 $y=(x-1)(x-2)(x-3)$
6 $y=(1-x)(2-x)(3-x)$
7 $y=2x-3$
8 $y=(2x+5)(x-1)$
9 $y=(x+1)(1-2x)$
10 $y=x(1-2x)(1+x)$

Even if the parabola is given in its general form $y=ax^2+bx+c$, it is still possible to draw an accurate sketch without much calculation.

Consider $y=2x^2+5x-4$.

This is a parabola with vertex downwards since $a=2$ is positive. The intercept on the y-axis is -4 (substituting $x=0$). The intercepts on the x-axis are not easy to find since $2x^2+5x-4$ does not factorise. Therefore, to find the position of the parabola in relation to the x-axis we start to complete the square.

i.e. $$2x^2+5x-4=2\{x^2+\tfrac{5}{2}x-2\}$$
$$=2\{(x+\tfrac{5}{4})^2+\text{constant}\}$$

The constant term serves only to move the parabola vertically; the term 2 widens or narrows the parabola.

This parabola will have the same axis of symmetry as the parabola $y=(x+\frac{5}{4})^2$.

i.e. the line $x=-\dfrac{5}{4}$ or $x=-\dfrac{b}{2a}$

Thus the parabola is vertex downwards, line of symmetry $x=-\frac{5}{4}$ and intercept on y-axis, -4.

See Fig. 23.

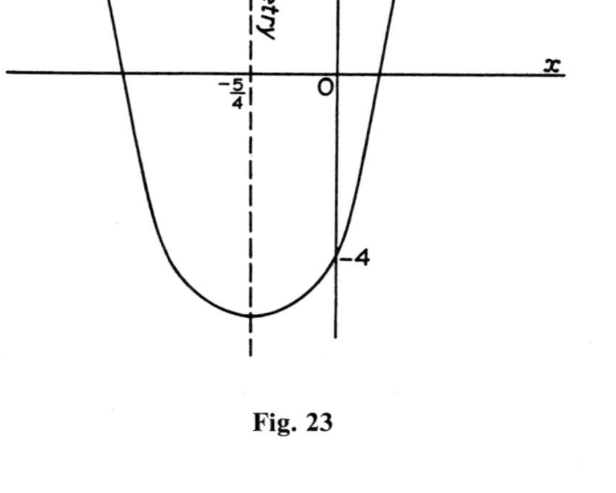

Fig. 23

Exercise 24

Sketch the following. In each case check these questions:

(i) Which way up?
(ii) Intercept on y-axis?
(iii) Intercepts on x-axes if possible?
(iv) Position of axis of symmetry $x=\frac{-b}{2a}$.

1 $y=x^2+2x-1$

2 $y=-x^2+4x-5$

3 $y=-2x^2+x-1$

4 $y=x^2-2x+3$

5 $y=x^2-3x-4$

6 $y=3x^2-6x+2$

7 $y=(2x-1)(x+2)$

8 $y=(1-x)(5+2x)$

9 $y=x^2-4$

10 $y=3x-x^2$

11 $y=4x^2+1$

12 $y=2x^2+5x$

VARIATION

Direct variation

The words 'y varies directly as x' imply that if x is doubled, then so is y; if x is multiplied by 3, then so will y be, etc.

Let us assume that x and y start with values 2 and 10.

Thus when $x=2$, $y=10$

when $x=4$, $y=20$

when $x=6$, $y=30$ etc.

Plotting these points gives the graph shown in Fig. 24.

You will see that this is a straight line through the origin with gradient 5.

i.e. $$y=5x$$

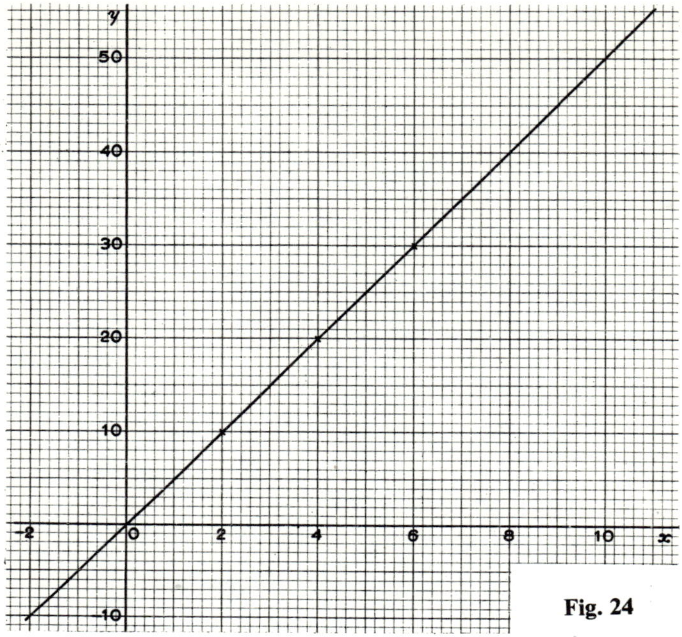

Fig. 24

Exercise 25

1 If x and y take the following pairs of values, plot a graph to find their relationship.

(a)

x	-3	-1	2	5
y	6	2	-4	-10

(b)

x	-2	0	4
y	-1	0	2

2 If y varies directly as x and $y=8$ when $x=2$, find the relationship between x and y and illustrate by a *sketch* graph.

Hint: substitute the values of x and y in the equation $y=kx$ to find k.

3 If y varies directly as x and $y=-5$ when $x=10$, sketch their relationship.

4 If $y=kx$, where k is a positive constant, sketch the relation between x and y.

5 If $y=kx^2$ (i.e. y varies directly as x^2), where k is a positive constant, sketch the relation between x and y then sketch the relation between x^2 and y.

6 If y varies directly as x^2 and $y=6$ when $x=2$, sketch the relationship between x and y and the relationship between x^2 and y.

7 If y varies directly as x^3 and $y=9$ when $x=1$, sketch the relationship between x and y. Also sketch the relationship between x^3 and y.

Inverse variation

y varies inversely as x implies $y=k/x$ where k is a constant.

Thus if x is doubled then y is divided by 2; if x is multiplied by a then y is divided by a. The graph of the relation between y and x in this case is clearly a hyperbola.

Exercise 26

1 If y varies inversely as x and $y=2$ when $x=2$, find and sketch the relationship between x and y.

2 If y varies inversely as x and $y=2$ when $x=-5$, sketch the relationship between x and y, and the relationship between $1/x$ and y.

3 If y varies inversely as x^2 and $y=4$ when $x=2$, sketch the relationship between x and y.

4 If y varies inversely as x^2 and $y=-2$ when $x=1$, sketch the relationship between x and y.

5 If y varies inversely as x^3 and $y=1$ when $x=-3$, sketch the relationship between x and y.

Partial variation

y is partly constant and partly varies as x implies $y=a+bx$ where a and b are constant. y partly varies as x and partly inversely as x^2 implies $y=ax+\dfrac{b}{x^2}$ where a and b are constant.

Joint variation

y varies jointly with x and z implies $y=kxz$ where k is constant.

y varies jointly with x^2 and the inverse of z implies $y=\frac{kx^2}{z}$ where k is constant.

Exercise 27

1 y is partly constant and partly varies as x. If $y=5$ when $x=1$ and $y=11$ when $x=4$, find the relationship between x and y and illustrate by a sketch graph.

2 y is partly constant and partly varies as x. If $y=3$ when $x=0$ and $y=1$ when $x=2$, sketch the relationship between x and y. Find y when $x=-2$.

3 y is partly constant and partly varies as x. If $y=-1$ when $x=3$ and $y=-3$ when $x=-3$, find y when $x=5$. Illustrate your answer by a sketch graph.

4 y varies inversely with x^2. If $y=8$ when $x=1$, sketch the relationship between x and y, and find y when $x=3$.

5 y varies partly as x^2 and partly as x. If $y=3$ when $x=1$ and $y=0$ when $x=0$, illustrate the relationship between x and y. What is the value of x which gives the least possible value of y?

6 y varies partly as x^2 and partly as x. If $y=0$ when $x=0$ and $y=-4$ when $x=2$, illustrate the relationship between x and y and state the value of x which gives the maximum value to y.

7 y varies directly as x^2. Illustrate by sketch graphs the various possibilities depending on the sign of the constant of variation.

8 The air resistance to a car is partly constant and partly varies as the square of the speed. If the resistance is 200 Newton when the speed is 50 km/h and 150 Newton when the speed is 20 km/h, find the resistance at 80 km/h.

Chapter Six

GRADIENT OF CURVE AND AREA UNDER CURVE

Gradient of a curve

Consider a straight line cutting a curve at a point P as in Fig. 25. Imagine this line to be pivoted at P and free to rotate. As it rotates there is one position where it does not cut across the curve at P (Fig. 25(b)). When the line occupies this position it is tangent to the curve at the point P. The gradient of this tangent is taken to be the gradient of the curve at P.

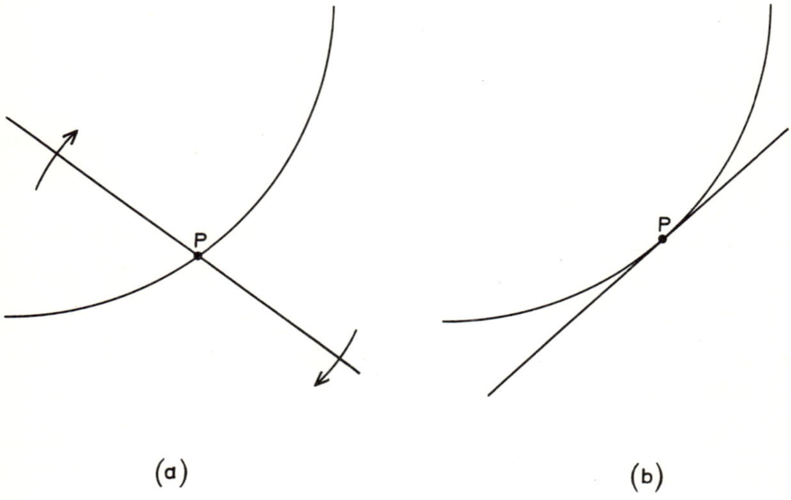

Fig. 25

Exercise 28

1 Draw the curve $y=x^2$ for $3 \leqslant x \leqslant 3$. Draw the tangents to the curve at the points $(-2, 4)$ and $(1, 1)$. Calculate the gradients of these tangents and hence give the gradient of $y=x^2$ at the points $(-2, 4)$ and $(1, 1)$.

2 Draw the curve $y=-2x^2+1$ for $3 \leqslant x \leqslant 4$. Draw the tangents to the curve at the points $(-1, -1)$, $(0, 1)$ and $(2, -7)$. What is the gradient of the curve at each of these points?

3 Draw the curve $y=x^3+2x^2-x-2$. What is the gradient of this curve at $(-2, 0)$ and $(1, 0)$? Note that the tangents to the curve at $(-2, 0)$ and $(1, 0)$ each cut the curve at a further point.

4 What is the gradient of $y=x^3$ at $(-2, 8)$ and $(0, 0)$? Give the equation of the tangent at $(-2, 8)$.

5 Give the gradient of $y=1/x$ at $(-4, -0{\cdot}25)$ and $(1, 1)$. Note that the gradient of $y=1/x$ is always negative. Give the equation of the tangent at $(1, 1)$.

6 A train starts from rest and the distances travelled at various times are given below.

Time in seconds	0	5	10	15	20	25	30	35	40
Distance in metres	0	15	30	50	65	90	110	140	165

Draw a distance–time graph (travel graph) and find the speed of the train at (a) 18 seconds and (b) 27 seconds as accurately as you can by drawing tangents.

7 A stone was thrown vertically upwards from the ground and its heights at various times are given in the following table.

Time in seconds	0	$\frac{1}{2}$	1	$1\frac{1}{2}$	2	$2\frac{1}{2}$	3	$3\frac{1}{2}$	4	$4\frac{1}{2}$	5
Height in metres	0	12	21	28	32	33	32	28	21	12	0

Draw a graph and find the velocity of the stone when its height is first 25 metres. What is its velocity when it is again at a height of 25 metres? Find the velocity after $1\frac{3}{4}$ seconds and $4\frac{1}{4}$ seconds. What is the velocity when the height of the stone is 33 metres?

8 A car travelling at 8 km/h takes 40 metres to stop after the brakes have been applied. The distances travelled at various times after the brakes were applied are shown in the following table.

Time in seconds	0	$\frac{1}{2}$	1	$1\frac{1}{2}$	2	$2\frac{1}{2}$
Distance in metres	0	13	23	32	39	40

Draw a graph and find the speed after travelling (a) 20 metres and (b) 35 metres.

The area under a curve

If the curve is a straight line, then the area can be divided into triangles, rectangles and/or trapezia and the area directly calculated. Otherwise, at this stage counting the squares on the graph paper is the most straightforward method.

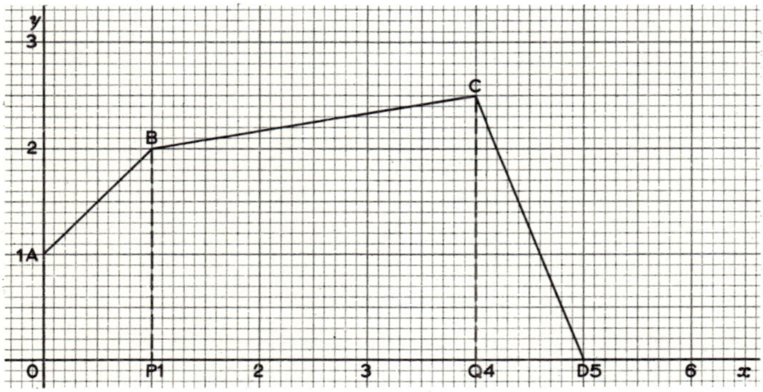

Fig. 26

In Fig. 26 the area $OABCDO$ has been divided into two trapezia, $OABP$ and $PBCQ$, and a triangle QCD.

Area $OABP = \frac{1}{2}(OA + PB) \times OP$ since OP is the distance between OA and PB

$= \frac{1}{2}(1 + 2) \times 1$

$= \frac{3}{2}$ square units

Area $PBCQ = \frac{1}{2}(PB + QC) \times PQ$

$= \frac{1}{2}(2 + 2\frac{1}{2}) \times 3$

$= \frac{9}{2} \times 3$

$= \frac{27}{2}$ square units

Area $QCD = \frac{1}{2}QC \times QD = \frac{1}{2} \times 2\frac{1}{2} \times 1$

$= \frac{5}{4}$ square units

$\therefore$ Total area $= \frac{3}{2} + \frac{27}{2} + \frac{5}{4}$ square units

$= 16\frac{1}{4}$ square units

Exercise 29

1 Find the area enclosed by $OABCDO$ where O is (0, 0), A(2, 1), B(3, 4), C(4, 6) and D(6, 0).

2 Find the area enclosed by $OABCO$ where O is (0, 0), A(2, −2), B(3, −5) and C(5, 0).

3 Find the area enclosed by $ABCDA$ where A is (−6, 0), B(−4, −5), C(−2, −3) and D(−1, 0).

4 A car accelerates uniformly from rest to 50 km/h in 15 seconds. It then travels for 10 seconds at 50 km/h before decelerating uniformly for 20 seconds and stopping. Draw a velocity time graph to illustrate this motion. Calculate the area under the curve (i.e. the area enclosed by your graph and the x-axis). On the velocity time graph the area under the curve represents the distance travelled. Thus the area which you calculated is the total distance travelled.

5 A train accelerates uniformly from rest to 60 km/h in 10 minutes. It then travels for 20 minutes at 60 km/h before decelerating uniformly for 15 minutes and stopping. Illustrate graphically, and use your graph to calculate the total distance travelled.

6 A particle starts from rest and moves in a straight line with a velocity of $(5t-t^2)$ metres per second at the end of t seconds. Find the distance travelled during the first 3 seconds of motion.

Hint: Draw the graph of $v=5t-t^2$ for $0 \leqslant t \leqslant 3$ and find the area under the curve by counting the squares on the graph paper.

7 A train starts from rest and moves with a velocity of $4t^2$ metres per second at the end of t seconds. Find how far it has travelled during the first 5 seconds of its motion.

8 A particle moves in a straight line with velocity (t^2+2) metres per second after t seconds. Find how far the particle travels during the third second of its motion.

Revision Exercise III

1 A manufacturer makes two types of radio-controlled hovercraft as toys—a standard and a de luxe version. The standard version takes 4 hours to assemble and 1 hour to finish whereas the de luxe version takes 5 hours to assemble and 2 hours to finish. The profits are £3 and £5 respectively. If the available time per week is 100 hours for assembly and 30 hours for finishing, how many of each should be made for the most profit?

2 Without using graph paper draw sketches of:

(a) $y=x$ (b) $y=2x-5$
(c) $y=x^2+1$ (d) $y=-x^3$
(e) $y=-\frac{1}{2}x+1$ (f) $y=2/x$
(g) $y=(x+1)(x+2)$ (h) $y=x^2+4x-3$
(i) $y=5$

3 Find the equation of the tangent $y=x^2+3$ at the point (1, 4).

4 The following are measurements of the speed of a lorry as it accelerates away from rest.

Time in seconds, t	0	1	2	3	4	5
Speed in km/h, s	0	2·1	4·0	5·9	8·0	9·9

Show there is an approximate linear law connecting s and t, and find this relationship.

5 Draw $\{(x, y): y=x^2+3x-1, -4 \leqslant x \leqslant 2\}$. Shade the area where $y > x^2+3x-1$. Give the minimum point of the curve and the co-ordinates of the points where it crosses $y=1$. Solve the equation $x^2+3x=1$.

6 Solve the following equations graphically:

(a) $y=-x^2+x+4$
(b) $y=2x^3+x+1$
(c) $y=\frac{3}{x}-5$

7 Illustrate $\{(x, y): y \geqslant x^2, 2 < y < 4, x \geqslant -1\}$.

8 Draw the graph of $y=2^x$ and find its gradient when $x=3$.

9 If y varies directly as x and $y=4$ when $x=1$, find the relationship between x and y, and illustrate by a sketch graph.

10 A particle starts from rest and moves in a straight line with a velocity of $(2t-t^2)$ metres per second at the end of t seconds. Find the distance travelled during the first 2 seconds of motion.

ANSWERS

Exercise 1

2 $y=-1$, a, e, f

Exercise 2

1 1 **2** 2 **3** $\frac{1}{9}$ **4** -2 **5** $-\frac{1}{2}$ **6** $\frac{1}{11}$

8 (a) 3 (b) -4 (c) 1 **9** (a) $\frac{2}{15}$ (b) $\frac{-3}{5}$ (c) $\frac{3}{5}$

Exercise 3

1 $y=x$; $y=2x$; $9y=x-3$
$y=-2x$; $y=-\frac{1}{2}x+3$; $11y=x+39$
$y=-2x+4$; $y=3x-3$; $y=-4x-2$
$y=x$; $15y=2x+20$; $y=-\frac{3}{5}x$
$5y=3x+10$

2 $y=-x+3$; $y=\frac{1}{20}x+2$; $y=\frac{-1}{6}x$
$y=\frac{1}{2}x-1$; $y=2$

Exercise 4

1 $m=6{\cdot}9$; $c=0$; $21\frac{1}{2}$ sec

2 $m=\frac{1}{5}$; $c=4$; 80

3 $m=10$; $c=3$; 1·2 sec; 2·4 sec; 3·7 sec; 5·7 sec; 103 m/s

4 $\theta=\frac{1}{2}t+15$

Exercise 5

1 $(1, -1)$; No; $x=1$

2 lowest point; $(-2{\cdot}7, 1)$ and $(0{\cdot}7, 1)$; Yes

3 highest point; $(1, 2)$, $(-1, -2)$ and $(3, -2)$, $(2{\cdot}4, 0)$ and $(-0{\cdot}4, 0)$. $x=1$

4 $(-1, -2)$, $(\frac{3}{4}, \frac{15}{8})$, $(1, 2)$ and $(4, 23)$

5 $4\frac{1}{2}$, $-\frac{1}{2}$; 0·25, 3·75; 1·3, 2·7

6 $x=2$ and $x=-4$, $(-1, -7)$

7 $x=1$ and $x=3$, touches $y=-2$ where $x=2$, $x=0{\cdot}8$ and $x=3{\cdot}7$, $y=-2$.

8 $y=9$, $x=5$ and $x=-1$, $x=-0{\cdot}2$ and $x=4{\cdot}2$, touches $y=9$ where $x=2$, $x=-1$ and $x=4$

9 $y=\frac{3}{4}$, $x=-\frac{1}{2}$, Yes

10 $x=-\frac{1}{4}$, $(0, 1)(-2, 5)(\frac{1}{2}, 0)$ lie on the parabola

11 $x=0$, $x=-0{\cdot}6$ and $x=1{\cdot}6$, $x=-1{\cdot}6$ and $x=0{\cdot}6$

12 $x=-\frac{1}{2}$ and $x=3$
Both minima $(0, 0)$ and $(0{\cdot}8, 5{\cdot}1)$

Exercise 6

1 $y=-3\frac{1}{4}$, $x=3{\cdot}3$ and $x=-0{\cdot}3$, $x=2{\cdot}6$ or $x=0{\cdot}4$, $x=3{\cdot}8$ and $x=-0{\cdot}8$

2 $(\frac{1}{2}, \frac{9}{4})$, $x=-1$ and $x=2$, $x=-0{\cdot}75$ and $x=2{\cdot}75$, $x=-1{\cdot}6$ and $x=0{\cdot}6$

3 $y=-\frac{1}{4}$, $x=0$ and $x=1$, $y=0$

4 $(0, 4)$ and $(1\frac{1}{2}, 4)$

5 $(0, -2)$ and $(3, -2)$

Exercise 7

1 $(-\frac{1}{4}, 3\frac{1}{8})$, $x=1$ and $x=-1\frac{1}{2}$

2 $y=\frac{9}{8}$ Curve does not cross $y=5x+5$

3 $x=0$ and $x=-2$

4 $(\frac{1}{2}, -4\frac{3}{4})$ maximum, $(-2, -6)$ minimum

5 $x=0$ and $x=-1$

6 $(-2{\cdot}35, 0{\cdot}5)$ and $(0{\cdot}85, 5{\cdot}75)$

7 $(\frac{-1}{6}, \frac{-25}{12})$, $x=-\frac{1}{6}$

8 $(0, 4)$

Exercise 8

1 $x=-0{\cdot}4$ and $x=2{\cdot}4$

2 $x=-2{\cdot}3$ and $x=1{\cdot}3$

3 $x=-1{\cdot}7$ and $x=1{\cdot}2$

4 $x=-2{\cdot}3$ and $x=1{\cdot}3$

5 $x=-1{\cdot}13$ and $x=1{\cdot}46$

6 $x=-0{\cdot}15$ and $x=-2{\cdot}18$

7 $x=-1{\cdot}5$ twice

8 $x=1{\cdot}2$ twice

9 $x=-2$ and $x=\frac{1}{3}$

10 $x=-0{\cdot}75$ and $x=0{\cdot}25$

11 $x=-0{\cdot}17$ and $x=0{\cdot}5$

12 $x=0{\cdot}5$ and $x=1$

Exercise 9

1 $x=-1{\cdot}55$ and $x=2{\cdot}55$

2 $x=-3{\cdot}45$ and $x=1{\cdot}45$

3 $x=0{\cdot}3$ and $x=1{\cdot}7$

4 $x=-1{\cdot}6$ and $x=2{\cdot}1$

5 $x=-0{\cdot}2$ and $x=1{\cdot}5$

6 $x=-1$ and $x=0{\cdot}67$

Exercise 10

1 $(\frac{1}{4}, -1\frac{1}{4})$, $x=-\frac{1}{2}$ and $x=1$. No, $x=1\frac{1}{2}$ and $x=-1$, $x=-\frac{1}{2}$ and $x=1$, $x=-\frac{1}{2}$ and $x=1$, $x=0$ and $x=\frac{1}{2}$. No solution.

2 Touches the x-axis at $x=1$, cuts y-axis at $y=1$. Minimum value, $y=0$.
$x=1$ twice, $x=1$ and $x=2$, $x=-1$ and $x=3$

3 $x=-1{\cdot}8$ and $x=0{\cdot}3$, $x=-1{\cdot}85$ and $x=1{\cdot}35$

4 $x=-0{\cdot}2$ and $x=4{\cdot}2$, $x=0{\cdot}4$ and $x=2{\cdot}6$

5 $\sqrt{5}=2{\cdot}22$, $x=-1{\cdot}7$ and $x=2{\cdot}7$

6 $\sqrt{3}=1{\cdot}73$, $\sqrt{8}=2{\cdot}82$, $\sqrt{10}=3{\cdot}17$
$x=-0{\cdot}6$ and $x=1{\cdot}6$
$x=\pm 1{\cdot}4$
$x=-3{\cdot}55$ and $x=0{\cdot}55$

7 $(\frac{1}{3}, -1\frac{1}{3})$, $x=-0{\cdot}3$ and $x=1{\cdot}3$, $3x^2-3x-1=0$

8 $x^2+5x-3=0$, $x=-6{\cdot}55$ and $x=0{\cdot}55$

9 $2x^2-7=0$. $x=\pm 1{\cdot}87$

10 $x=-4.65$ and $x=0{\cdot}65$
$x=-4{\cdot}45$ and $x=0{\cdot}45$
$x=-4{\cdot}25$ and $x=0{\cdot}25$
$x=-4$ and $x=0$
$x=-3{\cdot}75$ and $x=-0{\cdot}25$
$x=-3{\cdot}4$ and $x=-0{\cdot}6$
$x=-3$ and $x=-1$

Exercise 11

1, 2 and 3

$$\left.\begin{aligned} x^2+x+4&=0\\ x^2+x+1&=0\\ x^2-4x+6&=0\\ -3-2x-x^2&=0 \end{aligned}\right\}\ \text{do not have solutions}$$

Revision Exercise I

1 gradient $\frac{1}{3}$, intercepts $(0, \frac{-4}{3})$ and $(4, 0)$

2 (a) $y=\frac{-4}{3}x+5$ (b) $y=\frac{1}{2}x$ (c) $y=1$
(d) $y=x+1$ (e) $y=2x+2$

3 $\theta=\frac{1}{180}h+15$, 13·25°C, 11·6°C

4 $x=\frac{5}{2}$, $(\frac{5}{2}, \frac{-17}{4})$
$x=-0{\cdot}4$ and $x=5{\cdot}4$, $x=0{\cdot}4$ and $x=4{\cdot}6$, $x=2$ and $x=3$

5 $2x^2=3x-1$, $x=\frac{1}{2}$ and $x=1$

6 $x=-0{\cdot}6$ and $x=3{\cdot}6$, $x=0{\cdot}4$ and $x=-2{\cdot}4$. No solutions

7 $d=2{\cdot}9s-44$; 85 m

Exercise 12

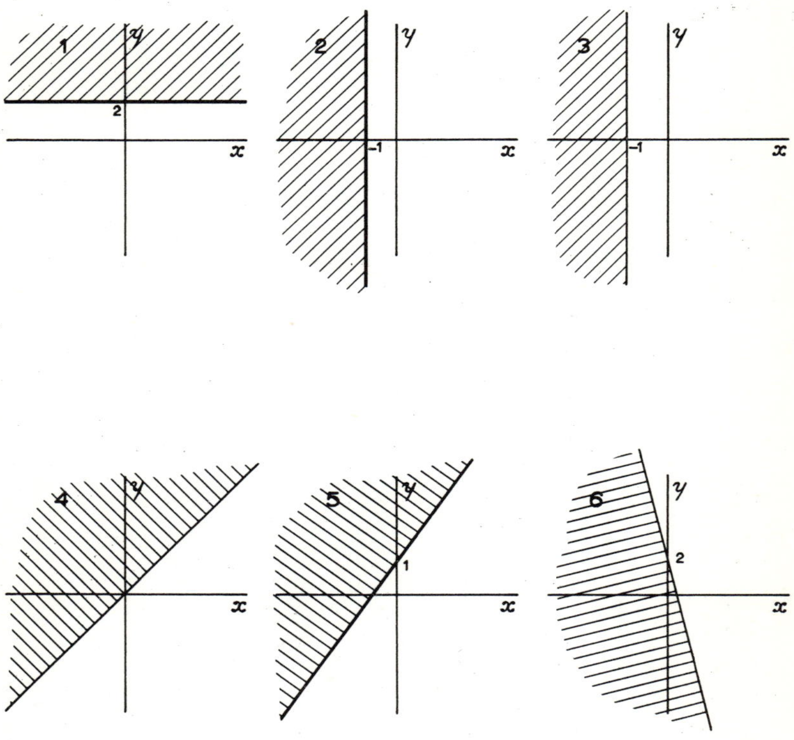

7
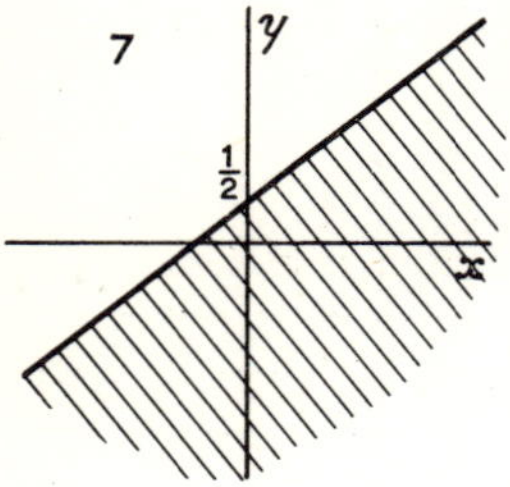

8
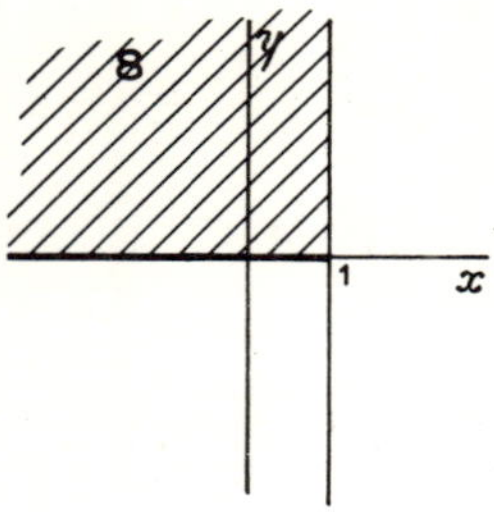

9
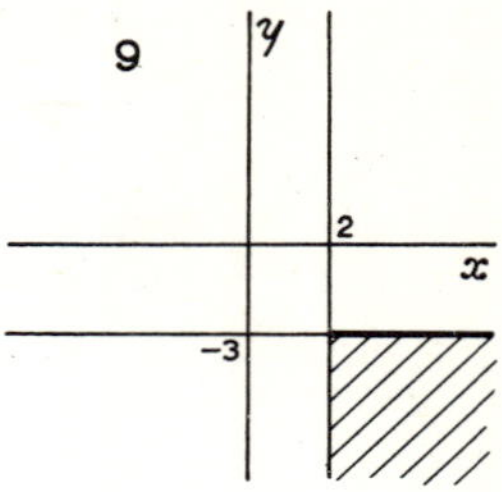

10
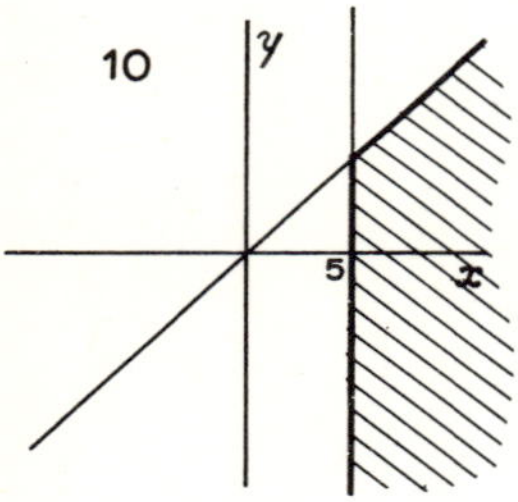

11
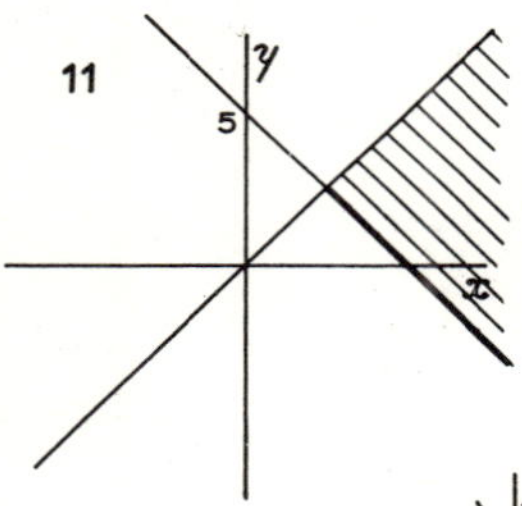

12
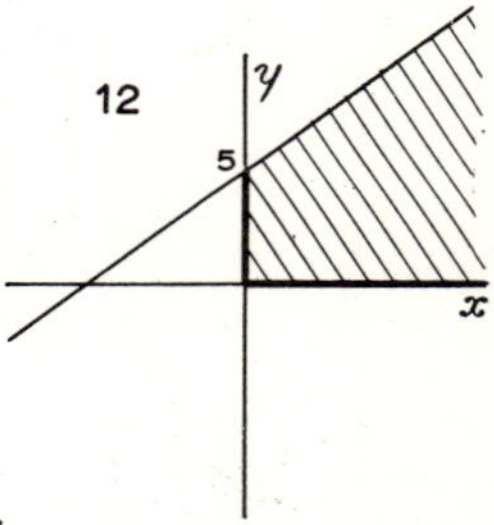

13
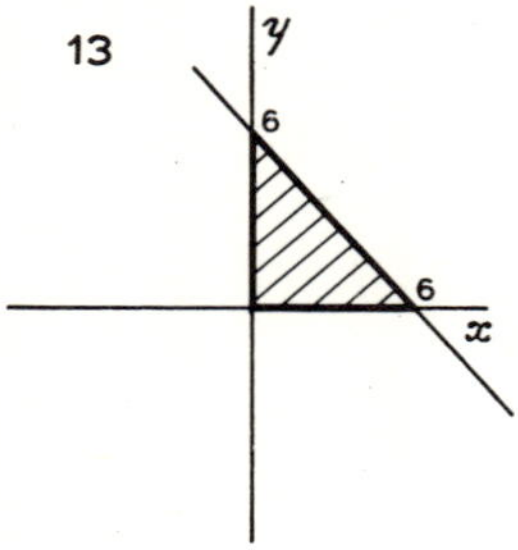

14
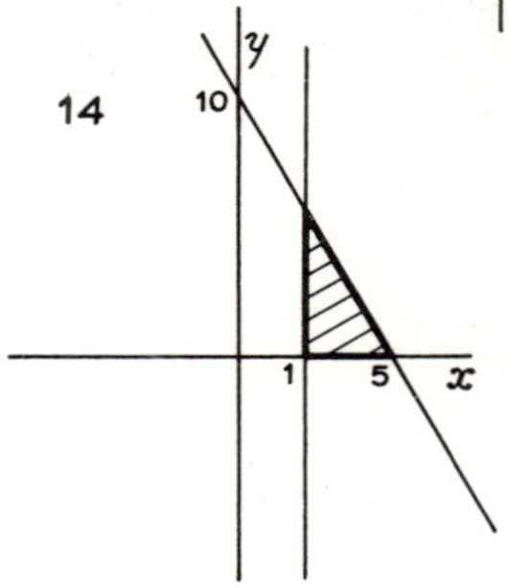

Exercise 13

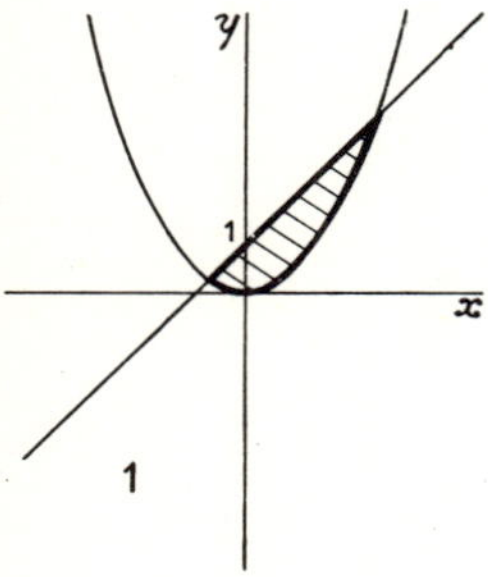

1

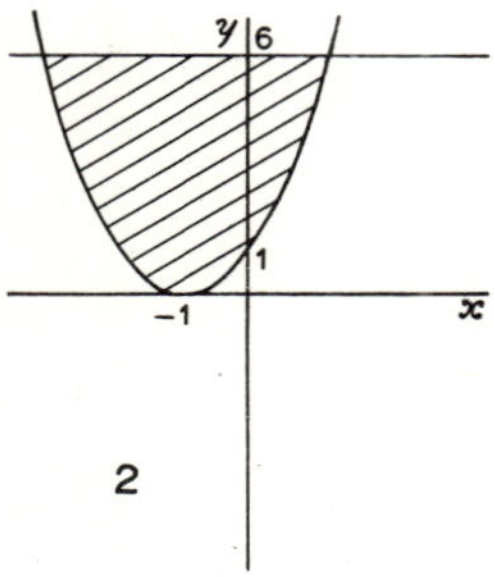

2

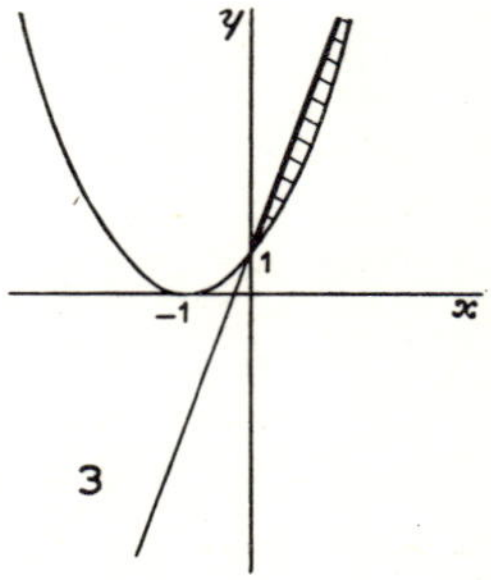

3

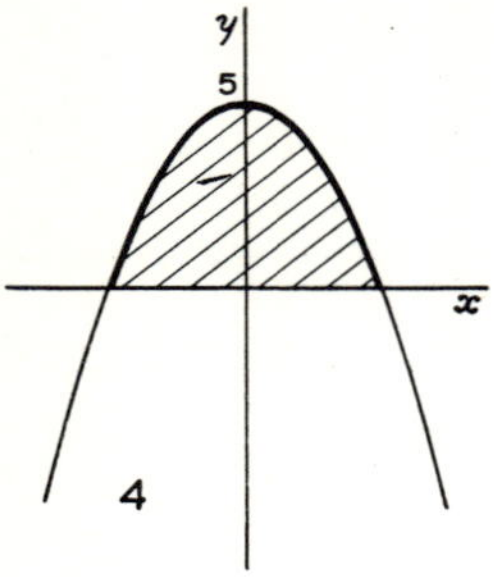

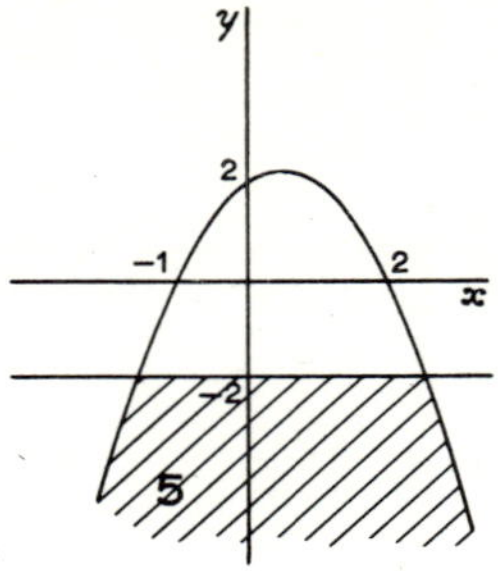

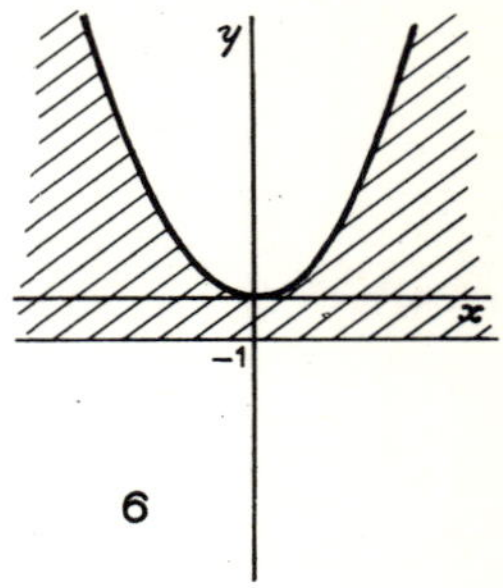

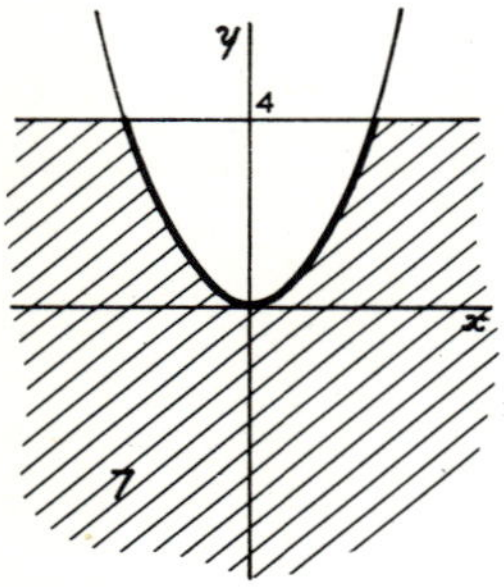

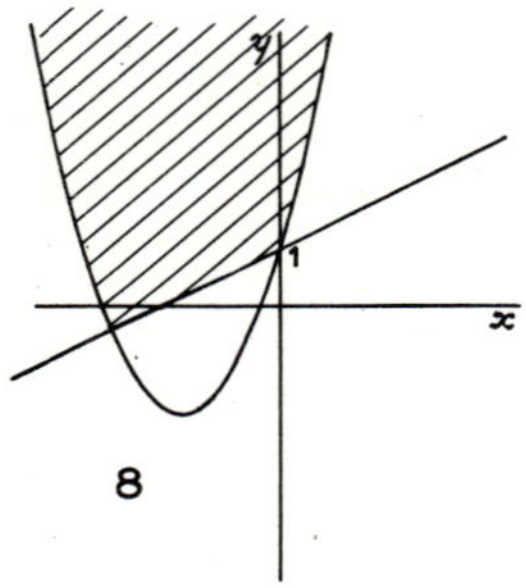

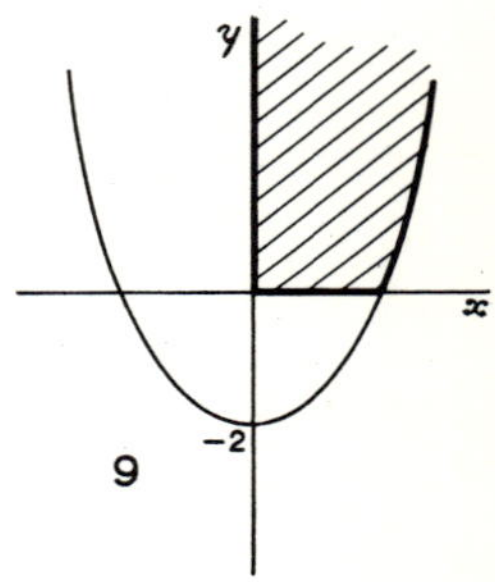

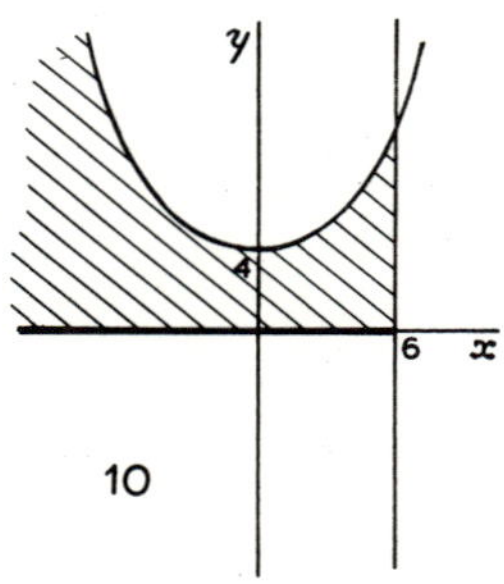

Exercise 14

1 30X, 60Y. £19·50

2 7A, 8B. £164

3 3 lb of A to 4 lb of B. £46·50

4 40 Type 1, 50 Type 2. £205

Exercise 15

3 3·7, 2·7, 3·3

Exercise 16

1 $x=0$ and $x=\pm 1$, $x=1{\cdot}5$

2 $x=1$ (twice) and $x=-1{\cdot}7$, $x=0$ and $x=\pm 1{\cdot}4$, (1, 0) minimum, (−1, 2) maximum

3 (−1·2, −2·1) minimum, $x=-2, 0, 1$

4 once, three times, once, three times

5 $x=-4{\cdot}5$, $x=0$ and $x=1$ twice

6 (0·3, −0·2) (−0·6, 0·6) $x=1$, 0 and $\frac{1}{2}$

7 $x=2$ only $x=\pm 1$ $x=1$ and $x=-1$, twice

Exercise 17

1 $x=0$ and $x=\pm 2$ **2** $x=-1, -2, -3$ **3** $x=\frac{1}{2}, 1, -1$

4 $x=\pm 1, -2$ **5** $x=0{\cdot}65$

Exercise 18

1 $x=-1$, $x=1{\cdot}26$ **2** $x=\frac{4}{3}$ **3** $x=\pm 1{\cdot}4$

Exercise 19

1 $x=0$

2 $x=0$, $x=1{\cdot}2$

Exercise 20

1 2·8, 5·6, 11·2, 23·0

3 1·04, 0·74, 0·05

Revision Exercise II

1

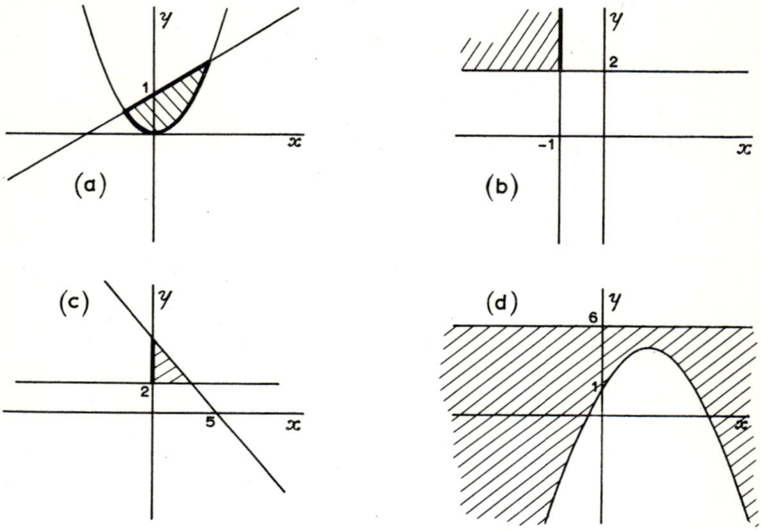

2 £880

3 (a) $y=\frac{1}{3}x+1$ (b) $y=x-6$ (c) $x=4$

4 $x=0{\cdot}3$, and $x=1{\cdot}7$. No solutions

5 $(1{\cdot}4,\ 3{\cdot}7)$, $(2{\cdot}6,\ -0{\cdot}4)$, $x=1$, $x=2$ and $x=3$

6 No maximum or minimum points. $x=1{\cdot}7$

Exercise 21

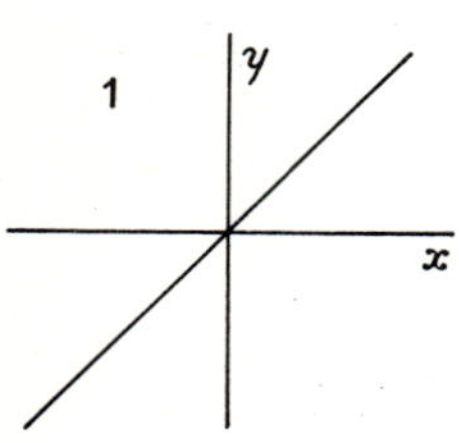

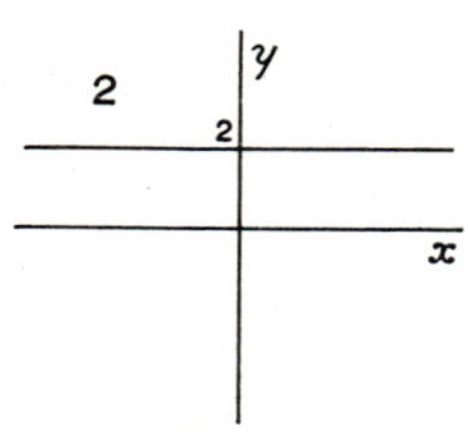

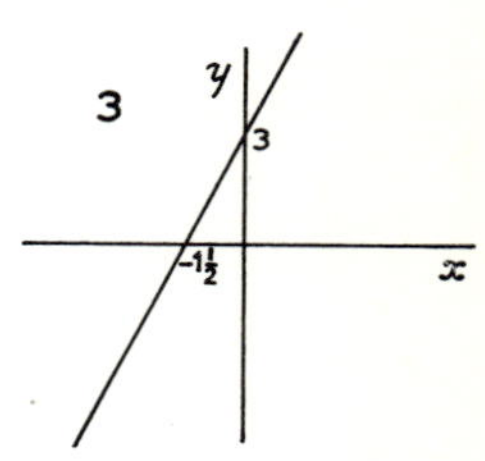

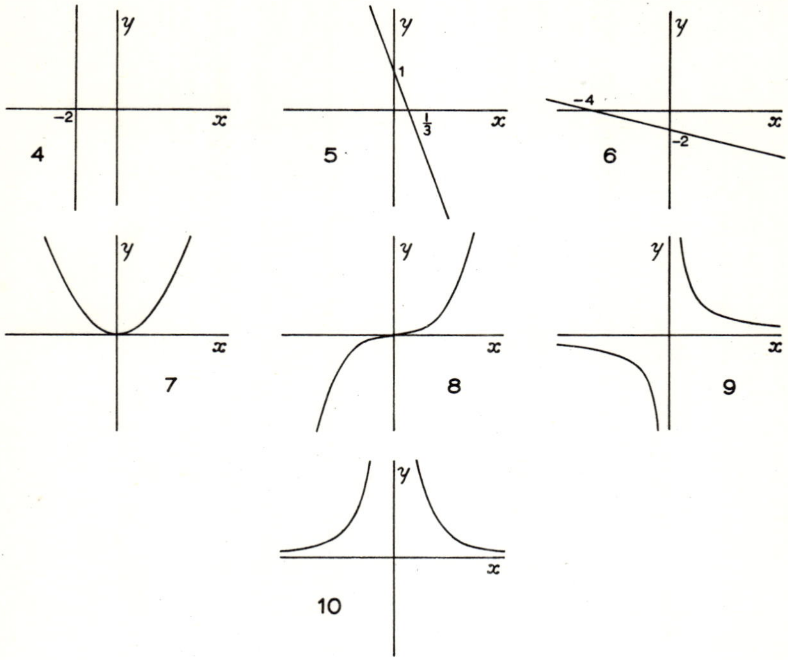
y
x
−2
4
1
1/3
5
−4
−2
6
7
8
9
10

Exercise 22

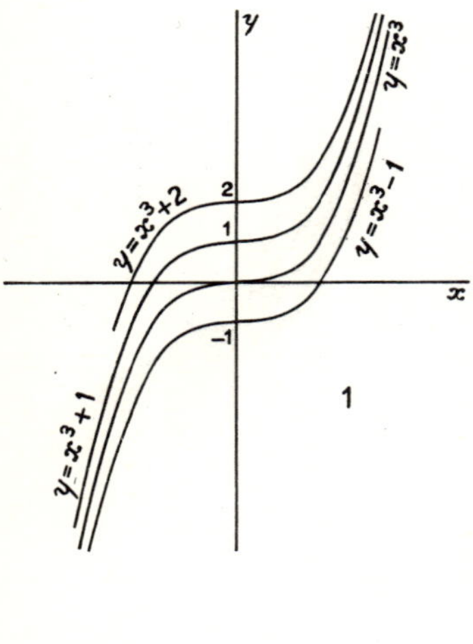
y
x
$y=x^3$
$y=x^3+2$
$y=x^3-1$
$y=x^3+1$
2
1
−1
1

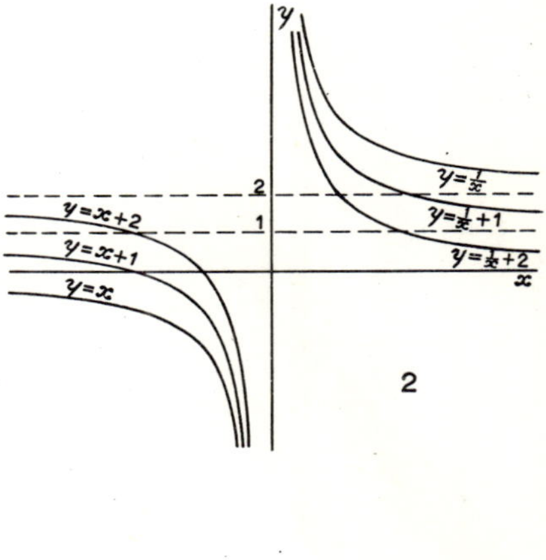
y
x
2
1
$y=x+2$
$y=x+1$
$y=x$
$y=\frac{1}{x}$
$y=\frac{1}{x}+1$
$y=\frac{1}{x}+2$
2

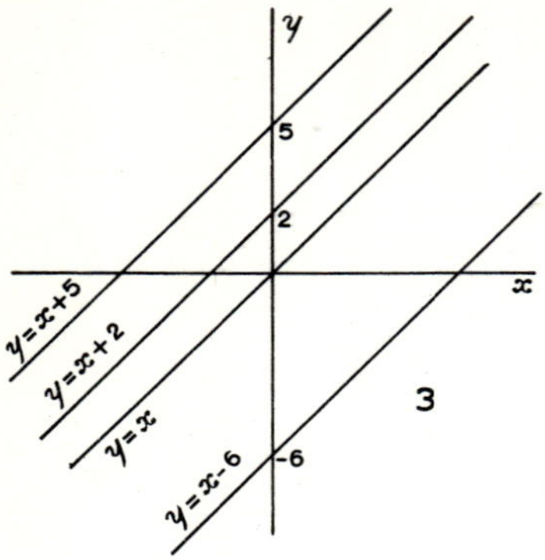
y
5
2
x
y=x+5
y=x+2
y=x
y=x-6
-6
3

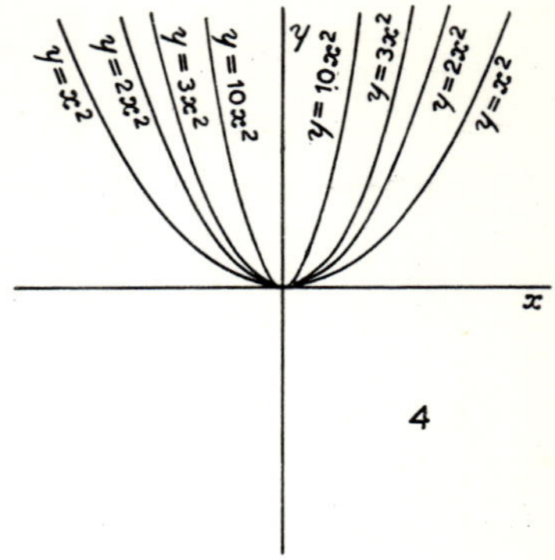
y=x²
y=2x²
y=3x²
y=10x²
y
y=10x²
y=3x²
y=2x²
y=x²
x
4

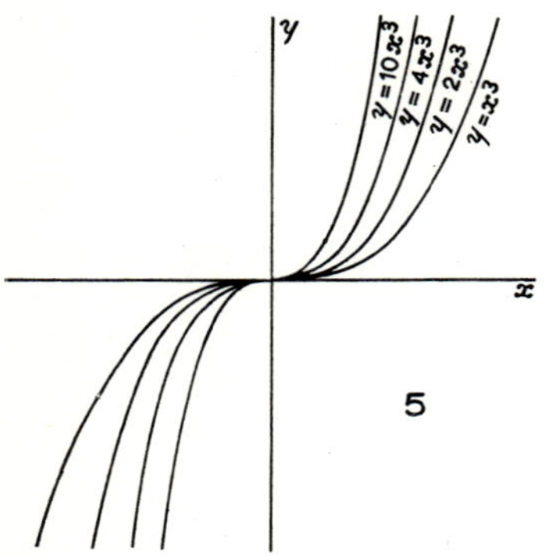
y
y=10x³
y=4x³
y=2x³
y=x³
x
5

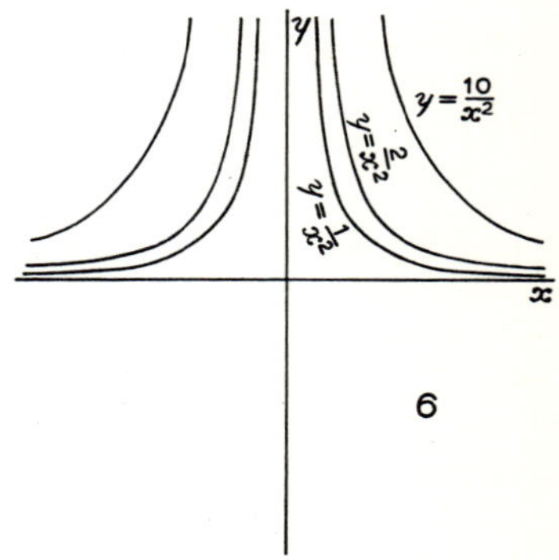
y
y=10/x²
y=2/x²
y=1/x²
x
6

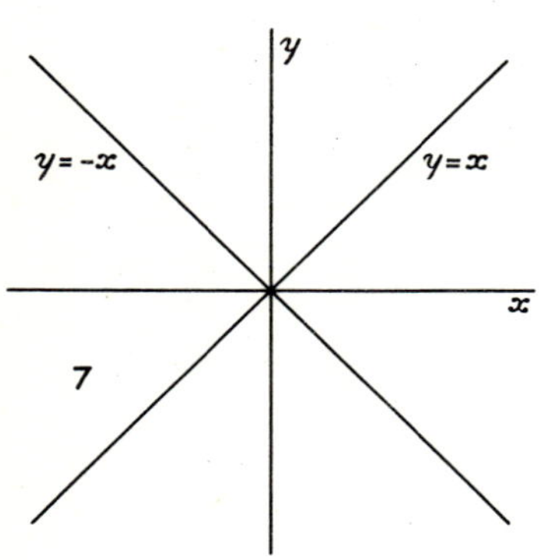
y
y=-x
y=x
x
7

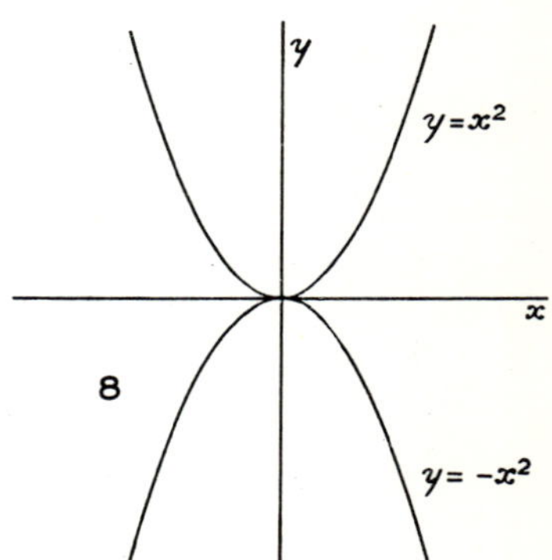
y
y=x²
x
8
y= -x²

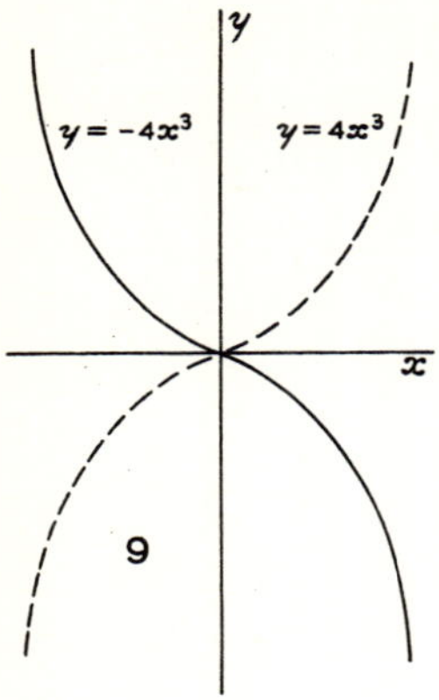
y
y = −4x³
y = 4x³
x
9

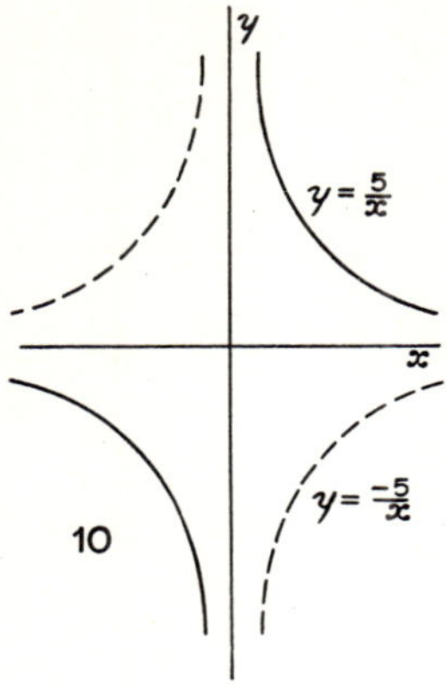
y
y = 5/x
x
y = −5/x
10

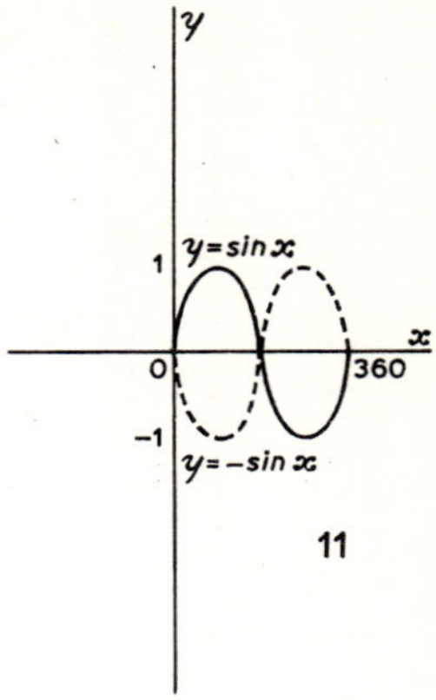
y
1
y = sin x
x
0
360
−1
y = −sin x
11

Exercise 23

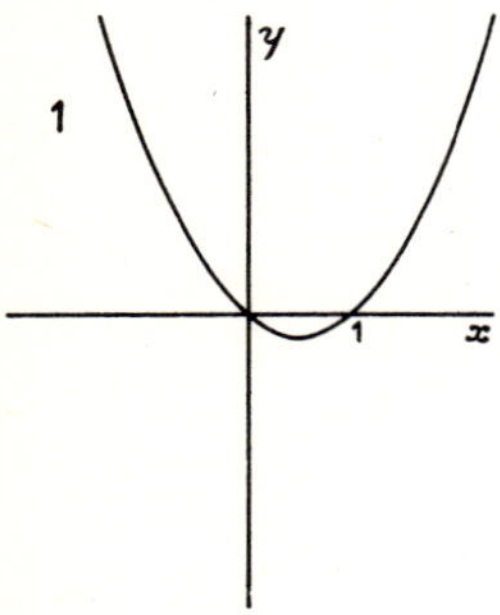
y
1
1
x

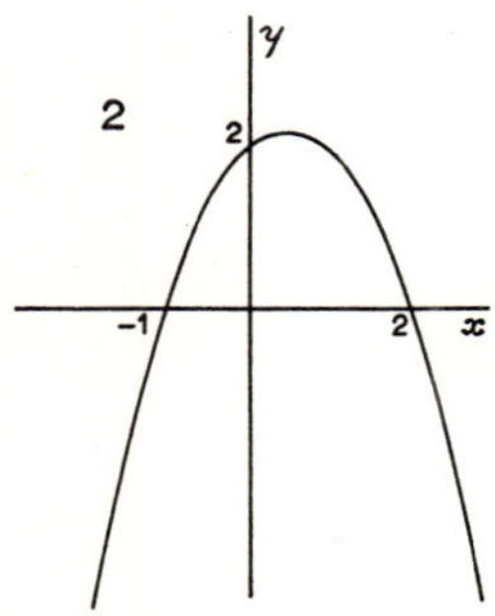
y
2
2
−1
2
x

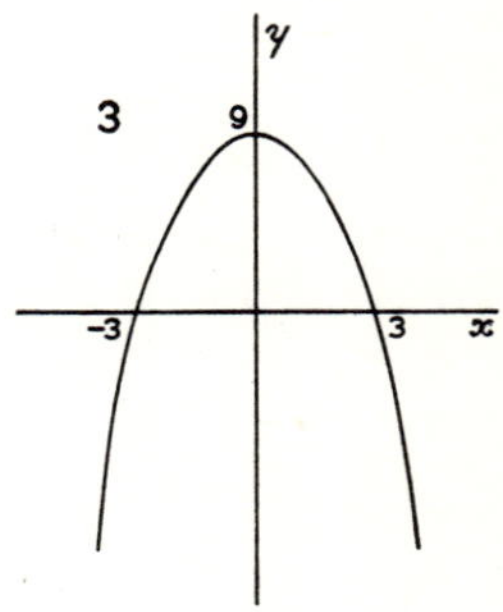
y
3
9
−3
3
x

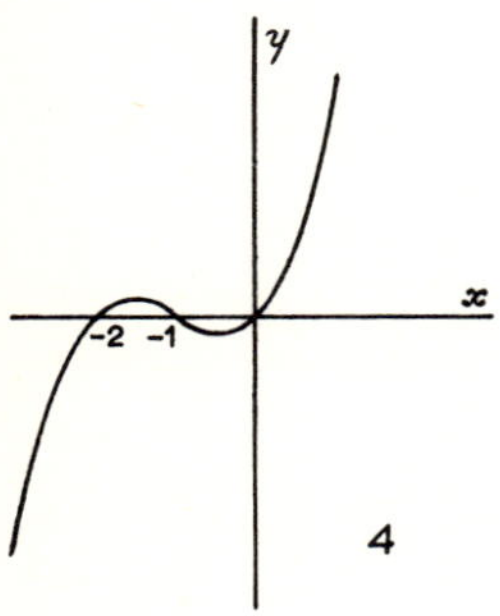
y
x
−2
−1
4

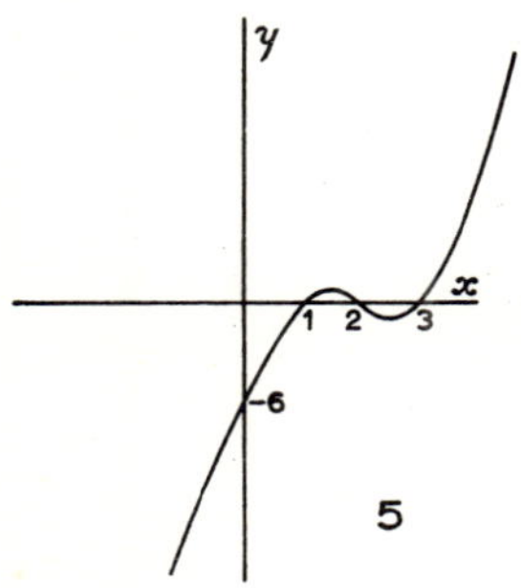
y
x
1
2
3
−6
5

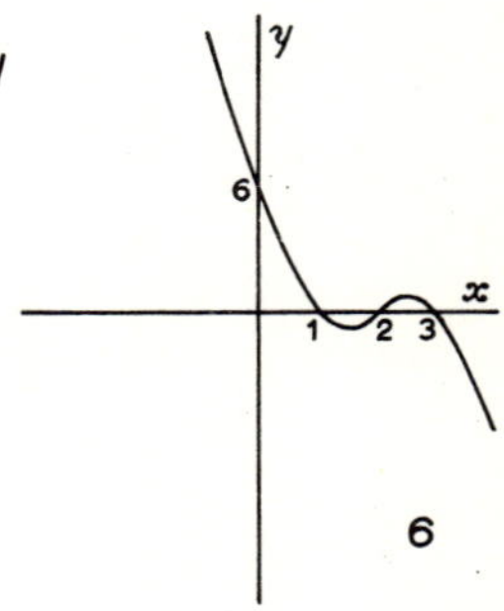
y
6
x
1
2
3
6

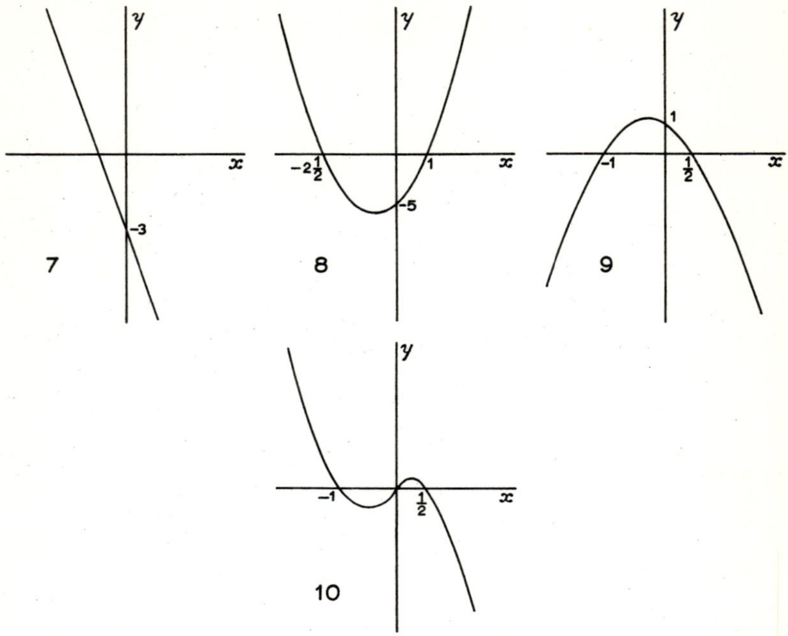

Exercise 24

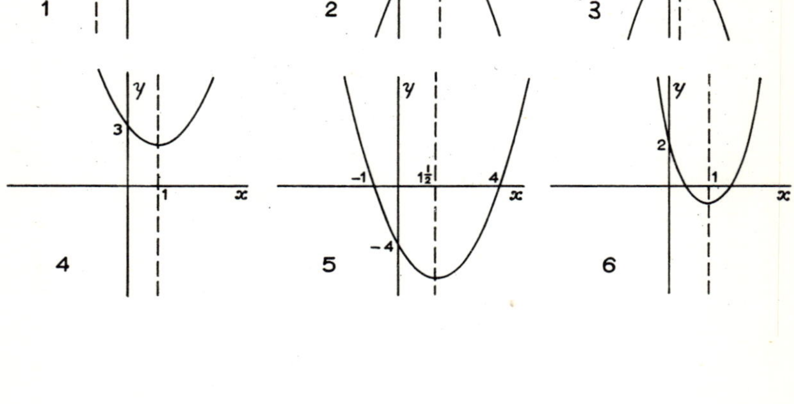

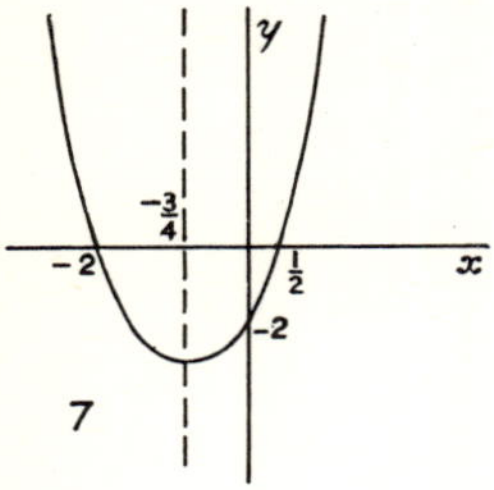

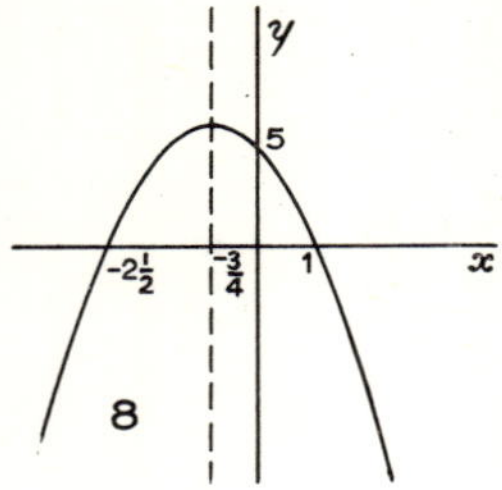

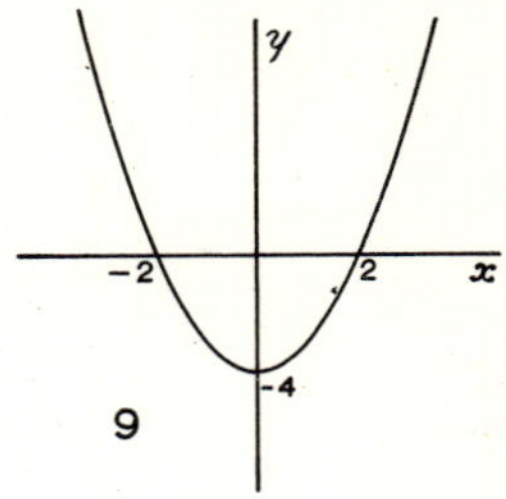

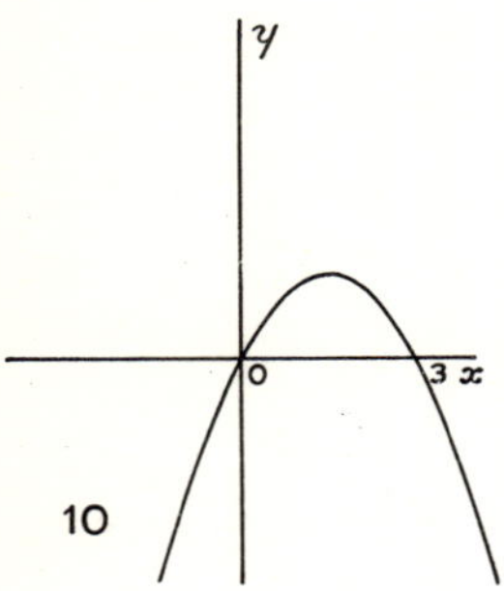

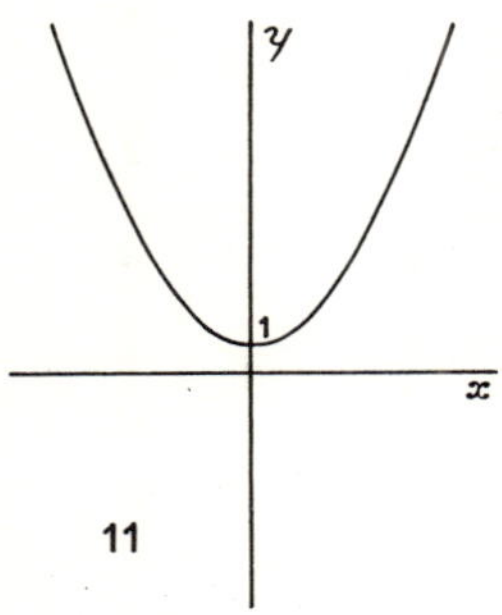

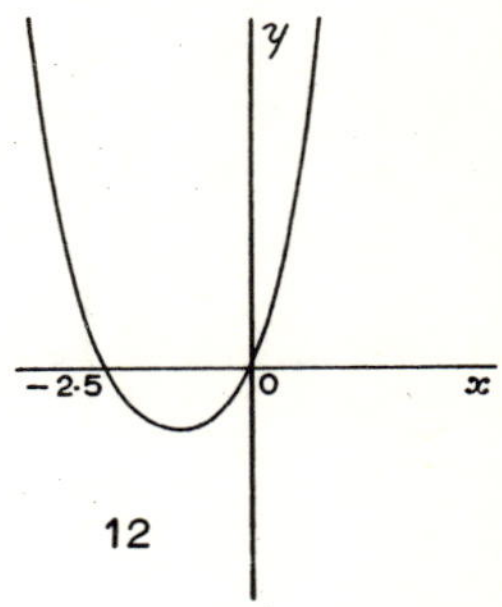

Exercise 25

1 $y=-2x$, $y=\frac{1}{2}x$

2 $y=4x$

3 $y=-\frac{1}{2}x$

6 $y=\frac{3}{2}x^2$

7 $y=9x^3$

(2, 3, 6, 7) For sketches see diagrams

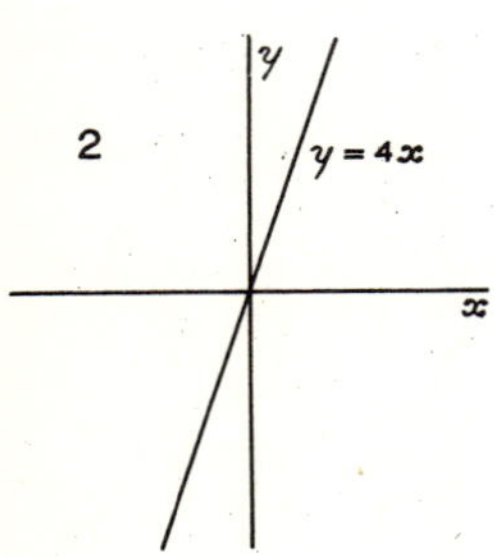

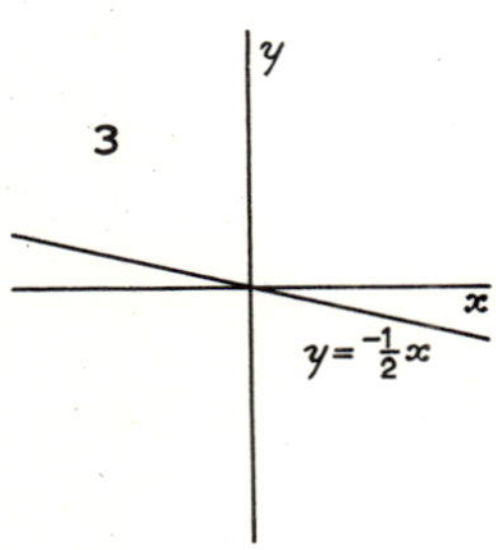

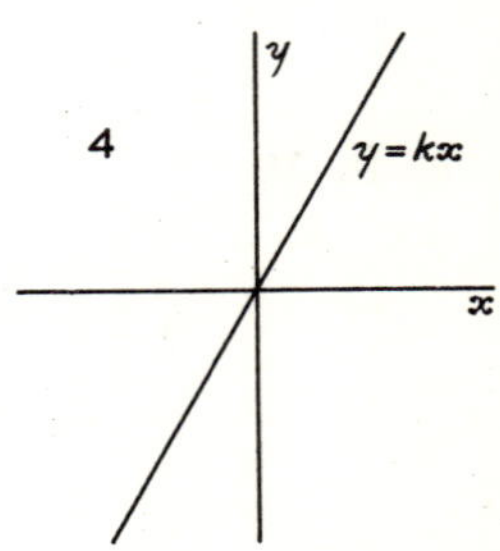

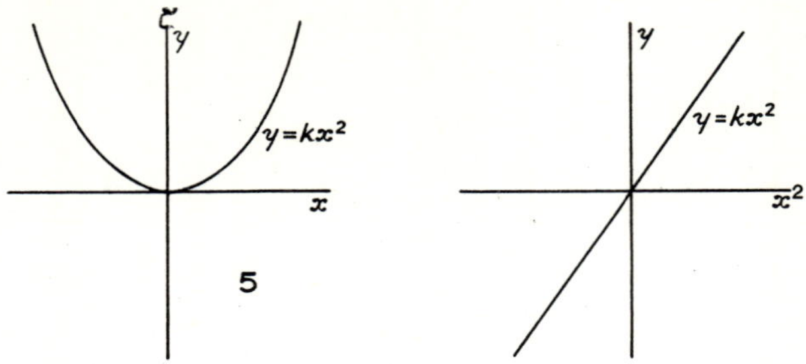

5

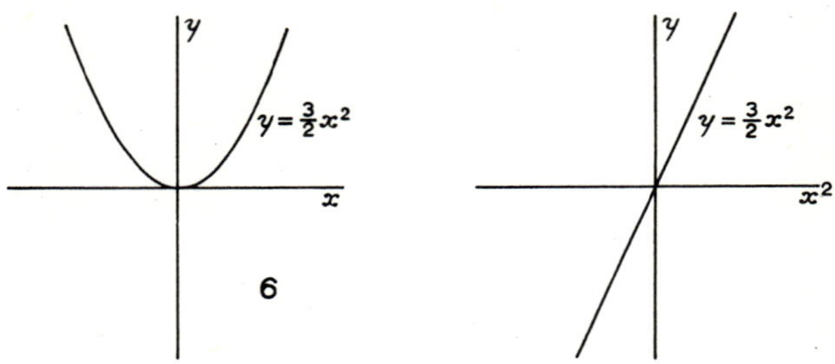

6

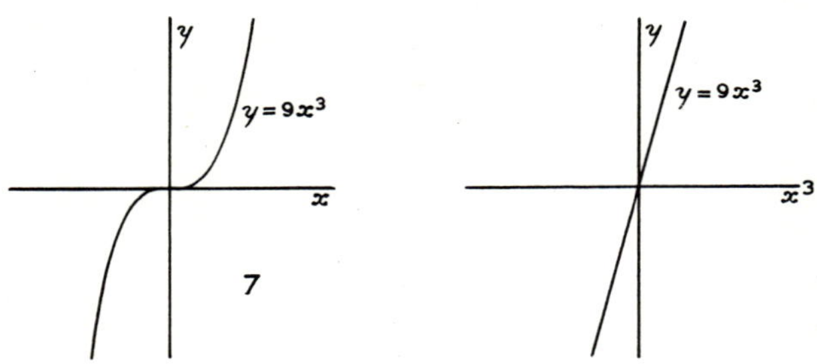

7

Exercise 26

1 $y=\frac{4}{x}$

2 $y=-\frac{10}{x}$

3 $y=\frac{1}{x^2}$

For sketches see diagrams

4 $y=-\frac{2}{x^2}$

5 $y=-\frac{27}{x^3}$

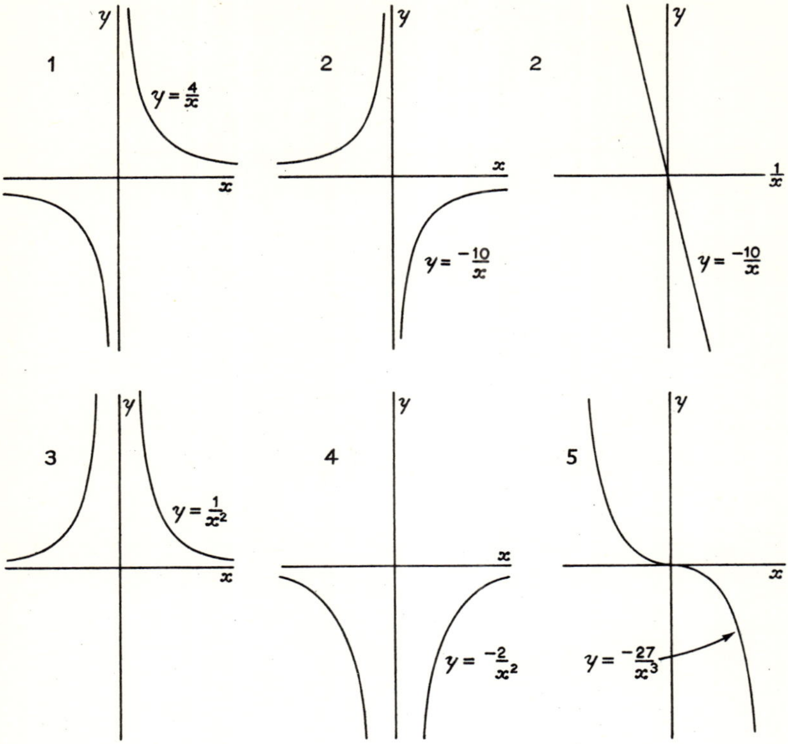

Exercise 27

1 $y=3+2x$

2 $y=3-x$, $y=5$

3 $y=\frac{x}{3}-2$, $y=-\frac{1}{3}$

4 $y=\frac{8}{x^2}$, $y=\frac{8}{9}$

5 $y=x^2+2x$, $x=-1$

6 $y=x-2x^2$, $x=\frac{1}{4}$

(1–6) For sketches see diagrams on page 78.

8 297 Newtons

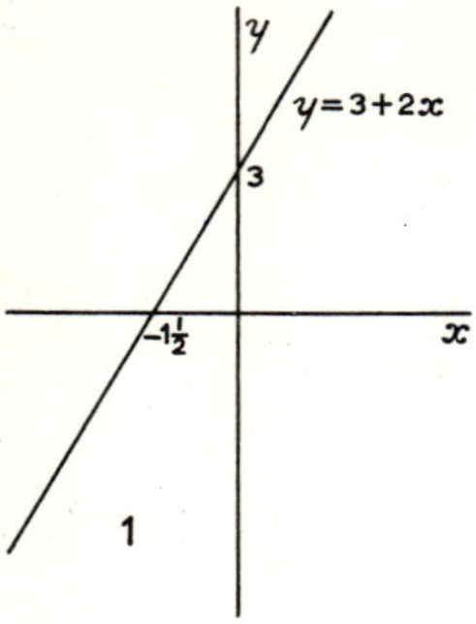

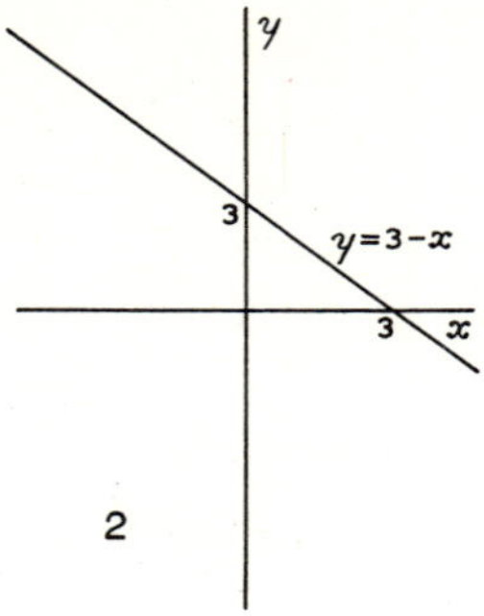

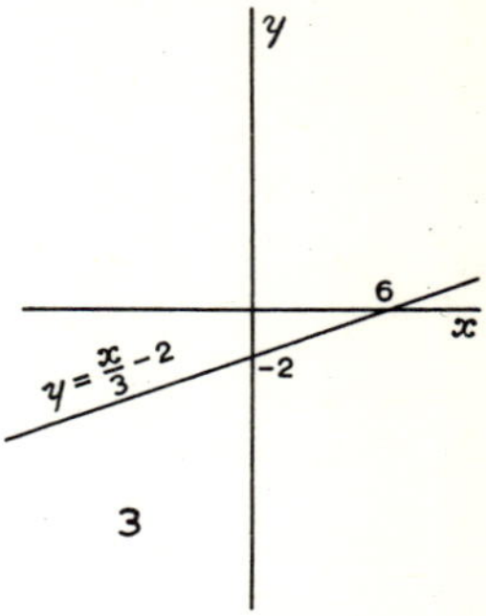

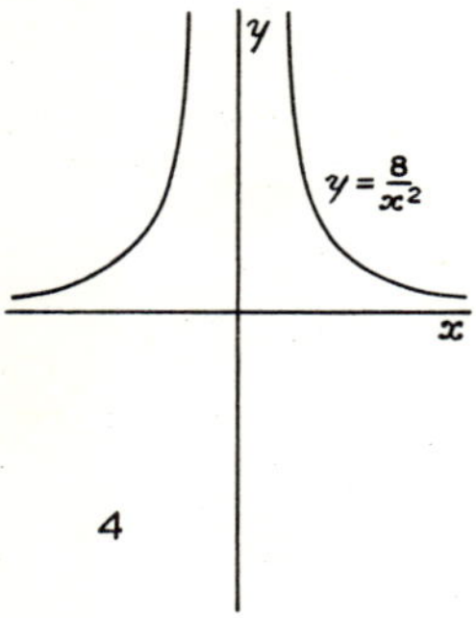

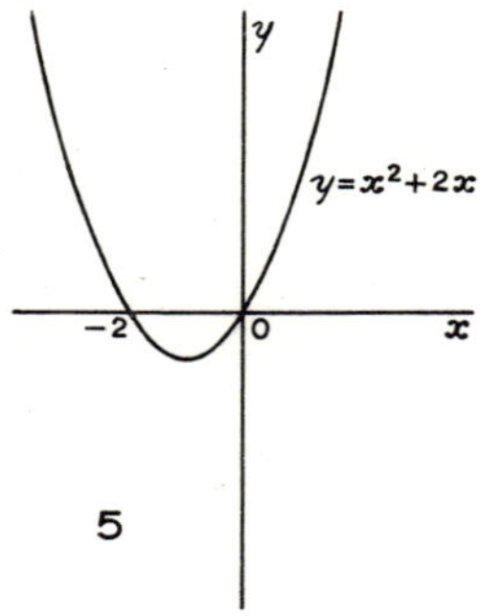

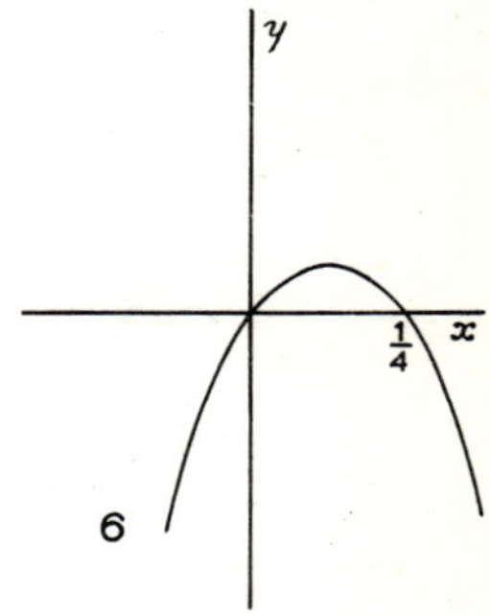

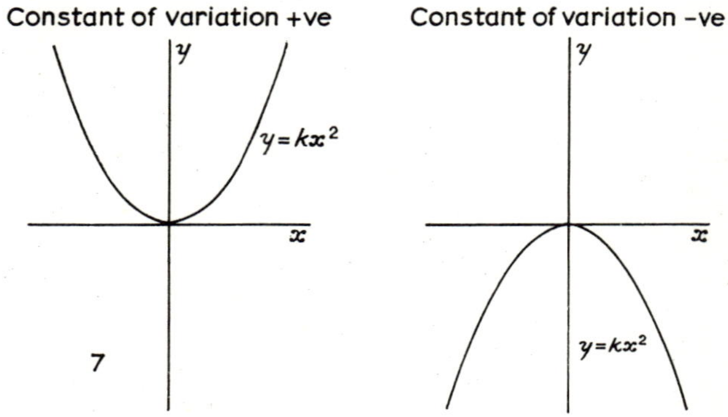

Exercise 28

1 $-4, 2$ **2** $4, 0, -8$

3 $-21, 6$ **4** $-12, 0.$ $y=-12x-16$

5 $-\frac{1}{16}$, -1. $y=-x+2$

6 40 m/s, 45 m/s

7 14 m/s upwards, 14 m/s downwards, 7·6 m/s, -18 m/s, zero

8 21 m/s, 13·3 m/s

Exercise 29

1 $14\frac{1}{2}$ sq units

2 $10\frac{1}{2}$ sq units

3 $14\frac{1}{2}$ sq units

4 $\frac{55}{144}$ kilometres

5 30 kilometres

6 $13\frac{1}{2}$ metres

7 $166\frac{2}{3}$ metres

8 $8\frac{1}{3}$ metres

Revision Exercise III

1 16 standard and 7 de luxe

2

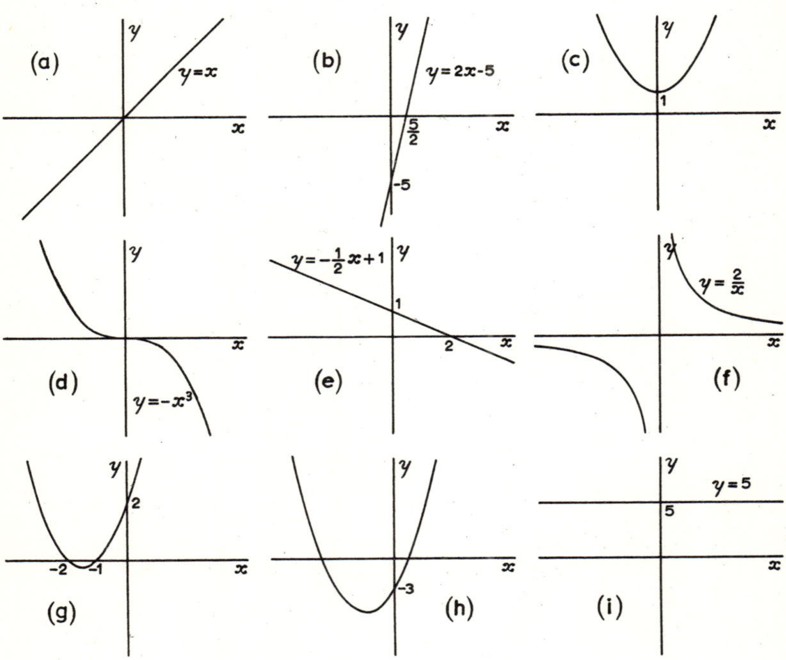

3 $y=2x+2$

4 $s=2t$

5 minimum $(-\frac{3}{2}, -\frac{13}{4})$, $x=0{\cdot}56$ and $x=-3{\cdot}56$, $x=0{\cdot}3$ and $x=-3{\cdot}3$

6 (a) $x=-1{\cdot}56$ and $x=2{\cdot}56$
(b) $x=-0{\cdot}56$
(c) $x=0{\cdot}6$

7

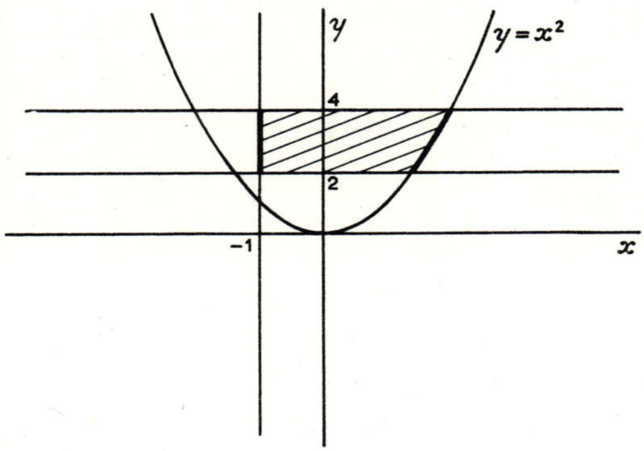

8 5.5

9 $y=4x$

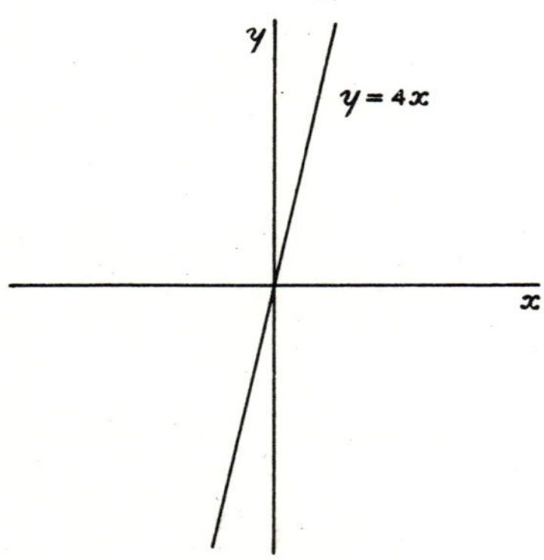

10 $1\frac{1}{3}$ metres